T0373589

Dance Lexicon in Shakespeare and His Contemporaries

This book provides a thorough analysis of terpsichorean lexis in Renaissance drama. Besides considering not only the Shakespearean canon but also the Bard's contemporaries (e.g., dramatists as John Marston and Ben Jonson among the most refined Renaissance dance aficionados), the originality of this volume is highlighted in both its methodology and structure.

As far as methods of analysis are concerned, corpora such as the VEP Early Modern Drama collection and EEBO, and corpus analysis tools such as #LancsBox are used in order to offer the widest range of examples possible from early modern plays and provide co-textual references for each dance. Examples from Renaissance playwrights are fundamental for the analysis of connotative meanings of the dances listed and their performative, poetic and metaphoric role in sixteenth- and seventeenth-century drama.

This study will be of great interest to Renaissance researchers, lexicographers and dance historians.

Fabio Ciambella is a Research fellow at Università della Tuscia, Viterbo, Italy.

Studies in Performance and Early Modern Drama
Series Editor: Helen Ostovich
McMaster University, Canada

This series presents original research on theatre histories and performance histories; the time period covered is from about 1500 to the early eighteenth century. Studies in which women's activities are a central feature of discussion are especially of interest; this may include women as financial or technical support (patrons, musicians, dancers, seamstresses, wig-makers) or house support staff (e.g., gatherers), rather than performance per se. We also welcome critiques of early modern drama that take into account the production values of the plays and rely on period records of performance.

Civic Performance
Pageantry and Entertainments in Early Modern London
Edited by J. Caitlin Finlayson and Amrita Sen

Strangeness in Jacobean Drama
Callan Davies

The Self-Centred Art
Ben Jonson's Parts in Performance
Jakub Boguszak

Dance Lexicon in Shakespeare and His Contemporaries
A Corpus-Based Approach
Fabio Ciambella

Shakespeare's Hobby-Horse and Early Modern Popular Culture
Natália Pikli

For more information about this series, please visit: https://www.routledge.com/Studies-in-Performance-and-Early-Modern-Drama/book-series/SPEMD

Dance Lexicon in Shakespeare and His Contemporaries

A Corpus-Based Approach

Fabio Ciambella

LONDON AND NEW YORK

First published 2021
by Routledge
2 Park Square, Milton Park, Abingdon, Oxon OX14 4RN

and by Routledge
605 Third Avenue, New York, NY 10158

Routledge is an imprint of the Taylor & Francis Group, an informa business

British Library Cataloguing-in-Publication Data
A catalogue record for this book is available from the British Library

Library of Congress Cataloging-in-Publication Data
A catalog record has been requested for this book

ISBN: 978-0-367-54047-0 (hbk)
ISBN: 978-0-367-54119-4 (pbk)
ISBN: 978-1-003-08768-7 (ebk)

Typeset in Times New Roman
by MPS Limited, Dehradun

Contents

Acknowledgements viii

Introduction 1

PART I
Dancing in early modern England: A historical overview 7

1 Continental and indigenous sources for early modern dances in England 9

Dancing at the Inns of Court 11
Diaries and annals 16
John Playford's The English Dancing Master *(1651) 19*

2 Salome vs David: The early modern *querelle* on dance between Neoplatonists and Puritans 23

3 Folk and courtly dances 33

4 From Elizabeth to Charles, through James: Dance and politics 38

References 44

PART II
Dance and/as language: State of the art and methodological issues 49

5 A language *for* dance: Dance as language 51

6 Language *about* dance: Matters of corpus and textual linguistics 58

7 Early modern English lexicography: Limiting the scope 61

8 Corpus selection and investigation 64
The VEP Early Modern Drama Collection 64
The #Lancsbox software and lexical analysis 67
Which words to look for 69

References 73

PART III
Analysis 77

A 79
Almain/Allemand(e) 79

B 82
Bergomask 82
Branle/Brawl 84

C 88
Canary 88
Cinquepace/Sinkapace 91
Coranto/Courante 93
Country Dance 96
Cushion Dance 99

D 101

Dance 101

G 103

Galliard 103

H 106

Hay 106
Horn(ile)pipe 108

J 111

Jig/Gig/Gigue 111

L 113

(La)volta 113

M 116

Maypole Dance 116
Measure 117
Moresca/Morisco 119
Morris Dance 120

P 124

Passamezzo 124
Pavan 126

R 128

Round(el)/Ring(let) 128

References 133

Conclusion 137
Index 139

Acknowledgements

"Shall I compare thee to a summer's day?" (W. Shakespeare, Sonnet 18, v.1). Definitely not, although my 2020 summer's days – and not only those – were completely devoted to the writing of this book, at the peak of the COVID-19 pandemic, which probably encouraged me to stay at home instead of sunbathing at the beach with my friends and volleyball teammates. For this reason, the first big "thank you" goes to them and to my family – my mum Marina, my dad Sandro, my (euphemistically) exuberant and talkative sister Monica, my couch potato brother Michele (aka Melindo) and my very much beloved granny Amalia. Each of them tried to distract me from my "mission" with promises of tasty carbonaras and roast chicken (my two favourite dishes), pizza, sushi rolls, BBQ ribs, beach volley matches, beach parties and gin and tonic. My love for them grows day by day – as does my post-lockdown ab fat.

My first "professional" thanks go to Professor Alessandra Petrina (University of Padua) who kindly assisted me with Skype meetings at the dawn of this project. She was probably the very first early modern scholar I contacted as soon as I signed the contract and her assistance was invaluable in dealing with issues of periodization, which she did with the reassuring smile that is her distinguishing feature.

My deep gratitude also goes to my dear friends and colleagues Camilla Caporicci (University of Perugia) and Emanuel Stelzer (University of Verona) for their help with the Routledge publication process (and much more), and Domenico Lovascio (University of Genoa) to whose witty advice I owe the title of this book: "Fabio, your book won't sell unless you insert 'Shakespeare' in the title!" The three of them and I are also part of an astonishing group of young scholars, all members of the Italian Association of Shakespearean and Early Modern Studies (IASEMS) – we like to be called, very humbly, IASEMS Wits. We even organized an online post-lockdown ERC workshop on 25–26 June 2020 to discuss our ongoing projects and research, which allowed me to present some sections of my book and receive their very welcome suggestions and impressions. The other IASEMS Wits who took part in this workshop and to whom I wish to express my heartfelt thanks are, in alphabetical order, Luca Baratta (University of Naples "Parthenope"), Michela Compagnoni (Roma Tre University), Michele de Benedictis (University of Cassino and Southern Latium), Alice Equestri (University of Sussex), Carmen Gallo ("Sapienza" University of Rome), Gabriella Infante (King's College London), Maria Elisa Montironi (Universities of Macerata and Urbino), Cristina Paravano (University of Milan), Ilaria Pernici (University of Perugia), Cristiano Ragni (University of Genoa), Beatrice Righetti (University of Padua), Silvia Spera (University of Salento), Elena Spinelli (University of St. Andrews), Allison Steenson (University of Padua), and Angelica Vedelago (University of Verona).

I would be remiss in not thanking my former PhD supervisor (now a dear friend), Professor Daniela Guardamagna, under whose "sagacious" (her term) wing my passion for and knowledge of the early modern English period was nurtured and grew. Under her supervision I began to research dance in sixteenth- and seventeenth-century English plays, which earned me a first prize for the best PhD dissertation of 2016 awarded by the Italian Association of English Studies (AIA). Moreover, her contribution to the revival of Thomas Middleton's canon in Italy deeply influenced my understanding of this Elizabethan and Jacobean playwright and consequently my analysis of the dances he inserted in his plays.

In addition, I am eager to express my gratitude to the two women who have been helping me the most with my past and present research and, I have no doubt, will continue to do so. The first is my precious linguist colleague and friend Valentina Piunno, from the University of – well, she actually has too many affiliations to choose only one! Everything I know about corpus linguistics is thanks to her ongoing patience at my incessant requests for help and opinions. I annoyed her every time I came up with an idea for this book and tortured her via Google Meet, Skype, and WhatsApp – even when she was desperately trying to get her son to sleep at midnight. One day I will no doubt have to look after him gratis while she goes on holiday with her husband.

Last, but definitely not least, this book is the result of the constant support of my – well, her precise role is difficult to classify. I refer to Professor Alba Graziano (Tuscia University of Viterbo), to whom I owe everything related to my career as a researcher of the English Language. And if while reading these lines she might think that her sometimes excessive pragmatism and impetuous character will prevent me from finding the right words to express my enormous gratitude towards her, this time I will prove her wrong. Working at her side while ever-new and amazing projects constantly pop out of her mind (all except one, and she knows which one) has truly enhanced my scholarly capabilities. It is an honour and a pleasure to work for and with her, and I truly hope our collaboration will last “till her retirement and beyond”, to quote the title of one of our latest efforts.

Introduction

This book is intended as a lexical and lexicographic guide aimed mainly at dance historians and those scholars interested in how dances were performed in early modern England and what connotations the different choreographies acquired in the plays by Shakespeare and his contemporaries. Although this project may not appear entirely original, given the lively scholarly debate surrounding Shakespearean dances, two methodological aspects guarantee its novelty. First of all, a larger picture of Elizabethan, Jacobean, and Caroline dances is depicted, thanks to the wide selection of a corpus of plays (666, as will be seen), one that sidelines the hegemonic position of the Bard's canon within sixteenth- and seventeenth-century theatrical output. In fact, as this book tries to demonstrate, Shakespeare was definitely not the only author who dealt with the polysemic values of dancing, nor was he the keenest mind when it came to terpsichorean knowledge. Nevertheless, in light of the number of studies devoted to the presence of dance in his canon (see, among others, Brissenden 1981/2001; McCulloch and Shaw 2019), their methodological and content issues are presented, adopted, and integrated herein.

The second and main methodological innovation of this book, however, is the corpus-driven analysis carried out using corpus linguistic software. A preliminary quantitative exploration of the dataset created is a paramount element of this research, since it would be unthinkable to explore the dance-related lemmas presented here, their collocational patterning and lexicosemantic neighbourhood without the aid of such software. Indeed, that software allows one to conduct very quick and precise targeted analyses on the lexicon of a selected corpus, which would in fact be quite a difficult task to accomplish with the human eye alone.

One should also clarify the label "early modern" from the outset. The debate around defining such periods as early modern, Renaissance, and Elizabethan is very lively nowadays, and thus confusion with labels is entirely plausible, especially following the advent of New Historicism. Among others, Petrina (2019) defends the idea that the label Renaissance can be used to describe Italian history but not English history, since "[h]istory and geography resoundingly clash" (150). In terms of the history of the English language, scholars usually agree that the early modern period extends from the mid-1400s (1476 is often taken as starting date, when William Caxton set the first printing press in Westminster) to the mid-1700s (more or less until 1755, when Samuel Johnson's *Dictionary* was published). Nevertheless, when it comes to English literature, early modern is often associated with the Tudor and early Stuart monarchs, conventionally from Henry VII's ascent to the English throne in 1485 to the outbreak of the Civil War in 1642. Yet, it is common to include the 18 years of the Wars, Interregnum, and Cromwell's Protectorate within early modern theatre, until the monarchic Restoration of Charles II in 1660, as plays continued to be secretly performed and published.

For the purposes of this work, however, I use the label "early modern"[1] with a restricted meaning to examine plays written and performed between 1576, when the first permanent Elizabethan public playhouse[2] was built in Shoreditch, London, to 1651, when John Playford published the first English dancing manual, *The English Dancing Master*.[3] This latter date has been chosen because it is highly symbolic for a dance-centred study about sixteenth- and seventeenth-century England. Indeed, Playford's manual is the first terpsichorean treatise ever written in English and the one which definitely attests to the national character of the English Country Dance, thus somehow acknowledging that the indigenous terpsichorean panorama was mature enough to break free from the hegemony of the Italian and French Renaissance dances which had entered Britain in the late 1500s. From Playford onwards, therefore, the English Country Dance becomes the emblem of the British national dance.

Part I of the book is devoted to a survey of the practice of dance in early modern England. Both continental and indigenous sources are considered and briefly introduced. On the one hand, Italian and French dance treatises such as Caroso's *Il ballarino*, Negri's *Le gratie d'amore*, and Arbeau's *Orchésographie* and their influence on the English terpsichorean environment are discussed; on the other hand, dance directions from the Inns of Court manuscripts are introduced as examples of indigenous sources for choreographic instruction. An entire section is dedicated to Playford's *The English Dancing Master*, given its pivotal contribution to the development of early modern English dance.

The short overview of sixteenth- and seventeenth-century manuals, treatises, and miscellaneous dance sources is followed by the main issues involving the terpsichorean practice and that contributed to its multiple metaphorical, allegorical, and symbolic connotations in early modern English plays. The debate between Neoplatonism and Puritanism regarding dance is introduced by considering the main voices of each faction; next, there follow some reflections on courtly vs folk dances. Both comparisons are then considered in light of pervading gender- and sexuality-related issues concerning the role of women within the dancing couples. In fact, if on the one hand both men and women contributed equally to the representation of the harmony of the spheres – according to English Neoplatonist devotees, and especially in solemn courtly dances – on the other hand the practice of dance was considered the most sinful of all, due to the dangerous proximity of the two dancing bodies and the sexual libido's arousal stemming from the lewd and lascivious movements of women's bodies – or so the Puritans maintained. Lastly, Part I considers the similarities and differences in aptitude towards the politics of dancing of the three monarchs who followed one another on the English throne during the period under scrutiny here: Elizabeth I Tudor, James I and Charles I Stuart.

Part II introduces a detailed reflection on the many connections between dance and language, drawing on a number of linguistic theories and different approaches. The main distinction presented in this part is between language *for* dance and language *about* dance, i.e., the difference between the various attempts to create a universal notation of dance with all its similarities with natural languages, and the study of writings on dance (treatises, manuals, specialised journals, etc.) through the contribution of various linguistic approaches (cognitivism, stylistics, corpus linguistics, etc.).

One chapter in this Part is dedicated to some of the most successful and interesting studies of early modern English lexicography in order to understand how a number of scholars have dealt with slippery issues related to the high variability – both lexicosemantic and morphosyntactic – of the English language in this period.

Indeed, drawing mainly on the expanding research field of Digital Humanities, the corpus selected for my investigation is presented only after a careful survey of the existing collections of early modern plays available. The Visualizing English Print (VEP) Early Modern Drama Collection project has been chosen as the corpus to be investigated, in particular its expanded version, which includes early modern plays until 1660. Following corpus selection, #Lancsbox is briefly introduced as the corpus linguistic tool chosen to carry out the corpus-driven analysis in Part III, and the sought for dance-related lexemes/lemmas are listed together.

The out-and-out lexicographic analysis is conducted in Part III. The dances in question, treated as keywords and nodes, are listed in alphabetical order, including a brief definition taken from authoritative dictionaries, after which a selection of the most interesting quotations from early modern plays is introduced. Next, the scenes quoted are commented on in reference to the collocational patterning and lexicosemantic neighbourhood of the lexeme/lemma searched for and the connotations each dance has in a particular scene or in all of them when they have something in common. The dance-related issues presented in Part I and the tools belonging to the Digital Humanities introduced in Part II are combined in Part III in order to analyse the many connotations that the various dances had in early modern English plays through the lenses of innovative and advanced computational tools.

The results provided by such an examination sometimes confirm and corroborate the hypotheses promoted by early modern dance scholars about the many connotations that the terpsichorean art had in sixteenth- and seventeenth-century England. Nevertheless, my study's lexical analysis and the exploration of the collocational patterning that dance-related terms have in the corpus selected sometimes clash with traditional views of terpsichorean criticism. However, this offers a far more varied panorama of early modern dancing, where continuous influences and confluences contributed to the variegated polysemy of dance in early modern English plays.

Notes

1 I will also distinguish, where necessary, among the Elizabethan (1558–1603), Jacobean (1603–25), and Caroline (1625–42) periods, according to the three monarchs who followed one another on the English throne in the years considered.

2 1567, when the Red Lion theatre was built, cannot be considered as a starting point for my research, since it was destroyed the following year and the only play ever performed there, as far as I know, was the anonymous lost play *The Story of Samson.*

3 At times, in Part III, the publication date of some of the plays is subsequent to 1651. Nevertheless, if the dates of (presumed or documented) first performance, entrance into the Stationer's Register, and/or first publication precede 1651, the play is considered within those analysed.

References

Brissenden, A. (2001) *Shakespeare and the Dance*. Second edition. Atlantic Highlands, NJ: Humanities Press.

Mcculloch, L. and Shaw, B. (eds.) (2019) *The Oxford Handbook of Shakespeare and Dance*. Oxford/New York: Oxford University Press.

Petrina, A. (2019) All Petrarch's fault: The idea of a Renaissance. *Memoria di Shakespeare: A Journal of Shakespearean Studies* 6, 145–164.

Part I

Dancing in early modern England: A historical overview

Dance is a tangible chronotopic manifestation of a well-defined culture in time and space. For this reason, as a cultural product, the terpsichorean art is particularly suited to exploring social, political, and ideological nuances of the multifaceted early modern English "world picture" – as E.M.W. Tillyard famously termed it (1943). Dance studies is a relatively new field of study,[1] which intermingles and benefits from a variety of research branches, such as cultural studies, performance studies, communicative studies, theatre studies, ethnology and anthropology (see Kaeppler 2000), gender studies, and linguistics, among others. Undoubtedly, a paramount contribution to the dissemination of dance studies was made by the so-called body or corporeal turn that, originating in philosophy and the neurosciences in the late 1960s and early 1970s, stresses the importance of the human body and corporeal semiotics in dance and dance performance.[2] Therefore, when such a complex and transdisciplinary field of studies is analysed from a diachronic perspective, taking into account a likewise complex and culturally rich period in the history of England, it is worth defining the object of research from the outset. In this first chapter, I present some pivotal aspects related to the practice of dance in early modern England, from philological issues that highlight the tension between indigenous and imported dances, to cultural aspects connected to the very complex sixteenth- and seventeenth-century ideological, religious, and political panorama.

First of all, it is necessary to review a series of indigenous as well as continental sources for late Medieval/early modern dances – be they extant manuscripts or printed writings – that directly or indirectly concern the practice of dance in England between the sixteenth and the seventeenth centuries. Next, gender-related issues are contextualised in the typical 1500s–1600s Neoplatonists vs Puritans *querelle*, thus considering arguments in favour of and contrary to dance, claimed by, respectively, Neoplatonic philosophers (and their English devotees) and the steadfast supporters and defenders of the Puritan ethics. Moreover, it is worth reflecting on the relationship between dance and social classes, thus exploring the distinction between court(ly) and folk/popular dances. Lastly, cultural and political issues concerning the practice of dance are explored with reference to the different or similar attitudes Elizabeth I, James I, and Charles I adopted under their respective reigns.

1 Continental and indigenous sources for early modern dances in England

As Emily F. Winerock (2011, 260) has pointed out, "there are no known English dancing manuals from the [early modern] period".[3] The scarcity of sixteenth-century British indigenous dance manuals is a long-standing issue that has forced scholars to dispute reliable reconstructions of how choreographies were performed during the early modern era in England. The only two manuals written during that period in Britain, before Playford's 1651 *The English Dancing Master*, were compiled in French by two French dancing masters, Barthélémy de Montagut and François de Lauze, who were both in the service of George Villiers, 1st Duke of Buckingham.[4] During the reign of James I Stuart, the two dancing masters published, respectively, *Louange de la danse* (1620–22) and *Apologie de la danse* (1623). Although de Montagut's treatise predates de Lauze's, in the preface to his *Apologie*, the latter claims that in 1620 he had shown his colleague a manuscript containing a draft of his manual and that his work was then plagiarized by de Montagut, who managed to publish *Louange de la danse* before de Lauze's treatise. Whatever the truth, it is undeniable that even at first glance one may easily notice that the structure of the two manuals and the dances recorded in them are almost the same.

De Lauze's work, considered the source text of Montagut's plagiarism, is divided into two macro-sections, one for men and one for women, dedicated respectively to George Villiers and his wife Katherine Manners. The first part of the treatise describes the male steps and the most suitable music for dances, such as the Coranto, Branle, and Galliard, with an initial introduction dedicated to the bow that gentlemen must perform when inviting ladies to take part in a ball. The second section describes women's steps for the same choreographies, plus directions of the Gavotte.

Despite the scarcity of indigenous English treatises, there is a rich tradition of dance manuals in sixteenth- and seventeenth-century continental Europe, where, especially in Italy and France, the art of dancing is industriously practised and theorized. The most authoritative dancing masters in Renaissance Europe were undoubtedly the Italian Marco Fabritio Caroso da Sermoneta and Cesare Negri, and the French Thoinot Arbeau. An overview of their works, presented below, shows a certain continuity and clear similarities between their choreographic directions and the dances described by de Montagut and de Lauze. Therefore, these intertextual references among Italian, French, and (pseudo-)English dance treatises seem to confirm that choreographies had not changed that much over a forty-year time span[5] in Europe, and that consequently even *Apologie de la danse* and *Louange de la danse* can be considered reliable sources for the study of dance in Elizabethan, Jacobean, and Caroline plays.

The first collection of Renaissance choreographies was Fabritio Caroso's *Il ballarino*, printed in Venice in 1581 and containing the names and steps of 77 dances, each accompanied by music scores and by a dedicatory poem to a well-known and wealthy Italian noblewoman. Caroso's treatise, as well as de Lauze's, is divided into two parts. It opens with a very detailed introduction on reverences and bows, then moves on to a list of dances and steps in the first part, and a series of poems in the second. *Il ballarino* was then revised and further supplemented by a second edition entitled *Nobiltà di dame*,[6] published once again in Venice in 1600 and including 'only' 49 dances, 20 of which are the same as *Il ballarino*, yet revised and corrected.

The French presbyter Thoinot Arbeau, an anagrammed pseudonym of Jehan Tabourot, published his dance manual *Orchésographie* in Langres, in north-eastern France, in 1589. This treatise is certainly less complete and complex than Caroso's *Il ballarino*, being less rich in technicalities and more discursive. Therefore, these characteristics of *Orchésographie* might well have facilitated its incredible spreading through Renaissance Europe (the second edition was issued in 1596), this manual being suitable also for non-expert dancers. The genre adopted by Arbeau is the Platonic dialogue between his fictional pupil, Capriol,[7] and Arbeau himself as a dancing master. Hence, this work does not adhere to a well-established tradition of conduct manuals – unlike its Italian counterparts – but turns out to be a small, discursive pedagogical volume about sixteenth-century dancing. *Orchésographie* opens with some general considerations on dance before analysing eleven dances – Basse Danse, Pavan, Galliard, Tourdion, (La) Volta, Coranto, Allemande, Branle, Moresca, Canary, and Les Bouffons – each with its own variations. Like Caroso before him, Arbeau also inserts musical scores to accompany the various steps during performances.

The last of these three great Renaissance dancing masters, Cesare Negri, printed his treatise *Le gratie d'amore* in 1602 in Milan, where he had been running a dance school since 1554. In 1604 he republished his manual with the new title *Nuove inventioni di balli*. The treatise, which owes much to Caroso's work, is divided into three parts. In the first, the author traces his own autobiographical profile as a dance scholar and teacher; in the second, he describes in detail the performance of the Galliard; and in the last, he provides information about other dances, with a section of music scores for lute. The only novelty of this treatise is its first-ever attested theorisation of the five basic positions of the feet in ballet.

The influence of continental dancing masters in the history of early modern dance in England is undeniable: one need only think about the Italian lemmas pertaining to the semantic sphere of dance that Florio translated into English in his 1598 Italian-English dictionary (reissued in 1611), or the fact that between 1608 and 1611 James I probably owned a copy of Caroso's *Il ballarino* purchased by the Royal Library (McManus 2002, 1).

Dancing at the Inns of Court

As already mentioned, there are unfortunately no dance manuals in English compiled by British choreographers or dancing masters. The few sources that deal with the terpsichorean practice in early modern England are sporadic hints in diaries or annals of the time (which will be dealt with in the next chapter) and a corpus of eight manuscripts,[8] the transcription of six of which appeared for the first time in 1965 by the scholar James P. Cunningham. Brissenden (2001, 6) lists the six documents transcribed by Cunningham in his preface to *Shakespeare and the Dance*. Nevertheless, in 1992 a seventh manuscript was discovered in Taunton, Somerset, which Brissenden surprisingly does not include in the 2001 revised edition of his book. Moreover, an eighth manuscript was officially transcribed in 2018 as a consequence of forensic evidence attesting that it was not a forgery.

The transcription of the first six manuscripts by Cunningham, however, presented various inaccuracies; hence, in 1986–87 the scholar David R. Wilson published a new and more accurate transcription, while the seventh MS was transcribed by James Stokes and Ingrid Brainard in 1992, and the eighth by Anne Daye and Jennifer Thorpe in 2018. It is worth noting that five out of eight extant manuscripts were certainly compiled by personalities who revolved around the London Inns of Court,[9] the four most famous and prestigious law schools in sixteenth- and seventeenth-century England – i.e., Lincoln's Inn, Middle Temple, Inner Temple, and Gray's Inn. One may wonder why personalities connected to the Inns of Court would write down lists of dances that were performed every year in the seasonal revels organised in the schools of law. Detailed information about these MSS will probably help better understand why the writers noted a variable number of choreographies (from a minimum of eight to a maximum of twenty-one), with surprising precision in the sequence of dances and steps, and with little variation from one MS to another. The eight MSS also identify a series of dances – eight to be exact – which always seem to recur on dancing evenings and usually in the same order of performance: the so-called Old Measures. Therefore, the revels at the Inns of Court would begin with this series of eight dances: Quadran Pavan, Turkeylone, Earl of Essex (or Earl of Essex Measure), Tinternell, Old Almain (i.e., Allemande), Queen's Almain, Madam Sosilia Almain (or Madam Cecilia Almain), and Black Almain.

The list of manuscripts, presumably in chronological order of composition, is as follows:

1. MS Rawlinson Poet. 108, ff. 10r–11r (Bodleian Library, Oxford). The manuscript belongs to Eliner Gunter (whose name appears on the title page), sister to Edward Gunter, a young man admitted to Lincoln's Inn in February 1563. The text can be dated around 1570 and contains love poems, songs, transcriptions of orations, and other scribbles. A total of fifteen dances are described. Given the different calligraphies that appear in the text, the drafting of the entire document cannot be attributable solely to Eliner, even though it is obvious that it was her brother Edward who could take part in the annual revels at the Inns of Court;
2. MS Dulwich College MSS, 2nd Series XCIV, fol. 28 (Dulwich College, London). According to Daye and Thorp (2018, 30–31), the possible dating of the manuscript ranges from c. 1570 to c. 1590, while the author is unknown. This MS was transcribed by John Payne Collier in 1844 for the first time, but it had always been considered forgery and did not receive the critics' attention, due to "Collier's tarnished reputation for forging and altering manuscript sources" (Daye and Thorp 2018, 27). It has been stored in the archives of Dulwich College since 1884 and when forensic evidence by Arthur and Janet Ing Freeman demonstrated that the MS is authentic, Daye and Thorp attempted a new, complete transcription. It lists the eight Old Measures, plus five "New Measures" whose tunes have connection with the Mulliner

Book, a sixteenth-century musical commonplace book written by Thomas Mulliner.

3. MS Harleian 367, ff. 178–179 (British Library, London). The author of the list of eight dances, which correspond exactly to the Old Measures, is unfortunately unknown. The annotations about choreographies and steps are part of a collection of notes and miscellaneous fragments written by the antiquarian John Stow(e) (d. 1605), but the handwriting used to list the eight Old Measures is completely different from the calligraphy of the author of the rest of the manuscript. The possible dating of the MS ranges from 1575 to 1625.
4. DD/WO 55/7, item 36 (Somerset Record Office, Taunton). The manuscript is signed by Io willoughbye (i.e., John Willoughby) and is the only text to record its year of composition: 1594. The circumstances of its discovery, in an environment far from the Inns of Court, together with the absence of any clear reference to one of the four schools of law, led the transcribers Stokes and Brainard to suppose that the dances described could also be performed at private parties in aristocratic country estates (1992, 1–2). Willoughby was never admitted to the Inns, but he had close contacts with those environments. In fact, in his collection of writings, there are twenty-two letters between him and two London lawyers who had studied at the Inns of Court: his stepson John Tuberbill and the brother of his son's second wife, William Davy of Creed. Perhaps Willoughby had taken part in one of the revels organised at the Inns of Court during a stay in London. However, at the time the manuscript was written, Willoughby was 23 and unmarried, and thus Stokes and Brainard's theory might be plausible.[10]
5. MS Douce 280, ff. 66av–66bv (Bodleian Library, Oxford). Again, the sequence of dances – twenty-one, the highest number of choreographies recorded in the eight manuscripts – is included in a miscellaneous list of notes, essays, and translations, this time with a precise pedagogical purpose: to educate a hypothetical preadolescent child. The author of the manuscript is J. Ramsey, admitted to the Middle Temple on 23 March 1605/6.
6. MS Rawlinson D. 864, 199r–199v (Bodleian Library, Oxford). Dated c. 1630–33, this manuscript is signed by Elias Ashmole, admitted to the Middle Temple only in 1657. The fact that the series of eight Old Measures had remained unchanged even some decades after the writing of the other four above-mentioned manuscripts suggests that the sequence of dances and their

choreographies performed at the revels had not changed. Ashmole (b. 1617), who was almost certainly less than twenty when he wrote down the series of eight dances, reports the name of the dancing master Rowland Osborne, who taught him how to perform the steps of the Old Measures. It should be noted, however, that when Ashmole compiled this part of the manuscript, he had not yet been admitted to the Inns, so it is difficult to define precisely the circumstances that led him to write the "copery of the old measures"[11] in f. 199 when he was still an adolescent. Between 1630 and 1633, Ashmole finished his studies at the Grammar School of Lichfield and studied law in London before becoming a lawyer 1638. The fact that the list of dances was compiled when Ashmole was probably still only approaching the Inns of Court could lead one to assume, as will be confirmed shortly, that there were places in England where dance was taught, i.e., dance schools.[12]

7. *Revels, Foundlings and Unclassified, Miscellaneous, Undated, etc.*, vol. 27, ff. 3r–6v (Inner Temple, London). The series of the eight Old Measures is followed by choreographic direction for the Cinquepace, the Argulius and other dances, and for the ensuing etiquette to open each choreography. Particular emphasis is given to the pivotal figure in any event, the Master of the Revels, known above all for being the person in charge of the censorship of the plays in the theatre, who also coordinated and directed the music and dances.[13] This is because the writer of the manuscript, dated c. 1640–75, is Butler Buggins, admitted to the Inner Temple in 1634 and Master of the Revels from 1672 to 1675. Although the writing of the MS took place from the Caroline era to the Civil War and the Restoration, in a profoundly changed context as compared to the Elizabethan and Jacobean periods, there is still continuity in the sequence of the eight main dances performed in the Inns of Court.
8. MS 1119, ff. 1r–2v and ff. 23r–24v (Royal College of Music, London). This manuscript basically contains a collection of songs. In ff. 1–2, in addition to the Old Measures, there is a precise description of the revel protocol with bows, performance, pauses, and a nine-verse composition entitled *An Holy Dance*. In ff. 23–24 there is a description of the steps of five other dances. The author is again Butler Buggins – although the calligraphy is very different from that of the MS mentioned above – who claims to have transcribed the steps and choreographies as they were taught by Robert Holeman, Master of the Revels before 1640.

With the sole exception of John Willoughby, critics agree that all the compilers are related to the Inns of Court environment. Moreover, the collocation of the lists of dances is the same in all eight manuscripts: i.e., the choreographies are reported among other miscellaneous texts as ordinary notes. It seems that reporting sequences of balls and steps was a sort of brain exercise assigned to would-be lawyers to improve their memory or, more likely, notes taken during dance lessons,[14] as Brissenden suggests (2001, 10).

There were many schools teaching how to dance properly in early modern London (see Eukbanks Winkler 2020). Not only would dancing masters teach their pupils the steps of the various choreographies, as Marchette Chute confirmed (1949, 89),[15] but they would also give pupils notions of correct body posture and ballroom etiquette to follow during dancing events. In addition, the (over)spreading of these institutions seems to be proved by the legal monopoly regulation introduced on 26 February 1574 by Elizabeth I and published in the *Calendar of the Patent Rolls*. In fact, in this document, the queen denounced the excessive increase in dance schools in London in the second half of the sixteenth century and imposed her royal monopoly on such businesses, which were clearly profitable.[16]

What has been asserted thus far concerns MSS nos. 1, 5 and 6, while the authors of the second and third manuscripts are anonymous and no. 4 was compiled by a person extraneous to the Inns of Court. Moreover, MSS 7 and 8, signed by the Master of the Revels Butler Buggins, probably had a very different purpose than the others, which may have been to pass on the traditional and well-established sequence of choreographies and steps prior to the outbreak of the Civil War. Whether they were merely a mnemonic writing exercise or an annotation for teaching/learning purposes, the eight manuscripts unequivocally link dance to the four most prestigious English schools for lawyers and provide us with information about the steps, the sequence of the Old Measures, and the etiquette to be followed during dancing events.

As is well-known, testimony of the time also confirms that the Inns of Court had a particular importance in the staging of some early modern plays that were presented in the schools of law (Hood Philips 2005, 23–36) – e.g., Shakespeare's *The Comedy of Errors* at Gray's Inn in 1594, *Twelfth Night* at the Middle Temple in 1602,[17] and probably also *Troilus and Cressida* in 1609 (see Elton 2000). Moreover, some of the early modern English playwrights were recorded in the Inns of Court registers. Among them, John Marston reports the greatest number of dances in his plays, specifying their names and directions precisely, unlike his contemporaries who inform their readers through "infamously brief" (Winerock 2011, 260) and rather vague stage directions, such as "a dance" or "they dance". This is the case, for instance, of the Galliard described twice in *Jack Drum's Entertainment* (5.1) and in *The Insatiate Countess* (2.1). Moreover, the French Branle is described and performed by Aurelia and Guerrino in the most famous of Marston's plays, *The Malcontent* (4.2).

Diaries and annals

In addition to the sources for courtly dances analysed thus far, other extant documents provide information about folk/popular balls in sixteenth- and seventeenth-century England; such texts are contained in miscellaneous works that, despite not focusing primarily on the art of dance, attest to its practice in this period.

The first choreographic instructions ever to appear in English are contained in a fifteenth-century manuscript housed in the Chapter Library of Salisbury Cathedral. This is the endpaper of an edition of Henry Medwall's interlude *Fulgens and Lucrece* (1512–16), in the second part of which a dance takes place. This endpaper describes the combination of forty-six steps and twenty choreographic variations of the Basse Danse, the most popular ball in the European courts between the late fifteenth and early sixteenth centuries.

In 1521, an appendix entitled *The Manner of Dancing of Basse Dances after the Use of France* appeared in Alexander Barclay's *The Introductory to Write and to Pronounce French*, a booklet about French pronunciation, orthography, and vocabulary, published by Robert Copland during Henry VIII's reign. The appendix contains seven Basse Danse choreographies. Its importance to the practice of dance is demonstrated not only by its appearance in such an early text, but also by the fact that it is the first evidence of the influence of continental dances – French in this case – on England.

In the introductory epistle to Barnaby Rich's *Barnaby Rich His Farewell to Military Profession* (1581), the English captain does not mention the Basse Danse among the forms he ironically asserts not to be able to dance. In fact, as Arbeau himself had declared in *Orchésographie*, "les basse-danses sont hors d'usage depuis quarante ou cinquante ans" (1589, online). *Farewell to Military Profession* – still considered one of the possible source texts of the dance-related dialogue between Sir Toby Belch and Sir Andrew Aguecheek in Shakespeare's *Twelfth Night* (1.3) – mentions not only court dances, such as the Galliard or the Cinque Pas, but also popular dances such as the Jig or the Hornpipe, which are not of French origin.

Almost twenty years later, the fashion for dances of French origin was still well-established in Elizabethan England.[18] In his *A Method for Travel* (1598), Sir Robert Dallington – writer, translator, and traveller, as well as director of the Charterhouse in London – wrote that "the French fashion of dancing is in most request with us" (cit. in Goose 2005, 116), complaining, however, about the influence of dances coming from Henry IV's Court and preferring dances of Italian origin. Dallington's complaint about choreographies from France in late sixteenth-century Britain underlines the increasing nationalism of Elizabeth I's reign, on the one hand, but, on the other, acknowledges the influence of Italian and French balls on the English terpsichorean panorama.

Sir George Buck (or Buc), Master of the Revels at James I's Court, offers an important example of how dance was taught at the Inns of Court. He wrote an appendix entitled *The Third University of England*, published with the 1615 edition of John Stow's *The Annales of England*. Buck's work is a small treatise cataloguing the subjects taught at the Inns of Court – the third higher education institution founded in England after Cambridge and Oxford. Among the many disciplines taught at the schools of law is the "Orchestice, or art of Dancing". The name of the subject, explains Buck, comes from the term ορχηστικη (orchestiche) or ορχησισ (orchesis), both used by the ancient Greeks to define the discipline of gymnastics, including dance, cubistics, and spheristics. The art of dance, writes Buck, recommended also by Plato as physical exercise for "ingenuous children", must be properly taught to gentlemen and used with due moderation not only in the "Courts of Princes" but also "in the best and most honorable Citties and euen in the colledges of the reuerend and graue Professors of our Lawes". In addition to providing information about the teaching of dance at Court and in the Inns of Court, Buck's text also attests to the spreading of the practice and teaching of dance in the main English cities, and to the existence of numerous books and pamphlets dedicated to the topic: "Of this art many books haue been written, as the Macaronicall tongue, intituled *Liber bragardissimus de danzis*, and diuers other the like". Unfortunately, the titles of these books and pamphlets are unknown, except for the small volume written in macaronic English entitled *Liber bragardissimus de danzis*, of which, however, there is no extant copy. Buck also quotes Tommaso Garzoni, a late-sixteenth-century Italian writer who had written *La piazza universale di tutte le professioni del mondo* (1585), a treatise whose forty-fifth chapter (or speech) was dedicated to dancers. This datum is very important because it provides information about the dissemination of Italian dance manuals in sixteenth- and seventeenth-century Britain.

Even in English pedagogical treatises, such as Roger Ascham's *The Schoolmaster* (1570) and Richard Mulcaster's *Elementary* (1582), dance is recommended as both a physical and a spiritual exercise to young gentlemen. In this regard, Mulcaster, an Anglican cleric and principal of the Merchant Taylors' School in London, stated:

> To keep things in order, there is in the soul of man but one, though a very honourable, way, which is the direction of reason. To bring things out of order there are two: the one strong headed, which is the commandment of courage, the other many-headed, which is the enticement of desires. Now dancing has properties to serve each of these: exercise for health, which reason ratifies; armour for agility, which courage commends; linking for allowance, which desire delights in. But because it yields most to delight (Cit. in Wagner 1997, 26).

Ascham, Queen Elizabeth's tutor, advises young gentlemen "to dance calmly, [...] in open places" (cit. in Wagner 1997, 25), and in daylight, never at night. The pedagogical aims of dance should therefore not be underestimated, especially in the light of what is contained in the earlier-mentioned manuscript Douce 280, wherein the author, Ramsey, addressed his directions regarding the terpsichorean exercise to a hypothetical son.

This brief overview of miscellaneous works dealing occasionally with the practice of dance closes with the autobiography of the eclectic Anglo-Welsh father of British deism, Lord Edward Herbert of Cherbury, a soldier, diplomat, poet, historian, and philosopher. In 1643, at the dawn of the Civil War, he collected the adventures and incidents of his life in a small volume that would be published for the first time only in 1886 by Sir Sidney Lee. Discussing his propensity for exercise while still a young man – during the reigns of Elizabeth, James, and, in part, Charles – Herbert confesses that he had excellent English, French, and Italian horse-riding teachers, yet he also liked to learn dance steps. Nevertheless, in his autobiography, he reports three simple choreographic exercises and then states that learning the right posture in dance

> gives one a good presence in and address to all companies, since it disposeth the limbs to a kind of *souplesse* (as the French call it) and agility, insomuch as they seem to have the use of their legs, arms, and bodies, more than any others, who, standing stiff and stark in their postures, seem as if they were taken in their joints, or had not the perfect use of their members. I speak not this yet as if

> it would have a youth never stand still in company, but only, when he hath occasion to stir, his motions may be comely and graceful, that he may learn to know how to come in and go out of a room where company is, how to take courtesies handsomely, according to the several degrees of persons he shall encounter, how to put off and hold his hat; all which, and many other things which become men, are taught by the more accurate dancing-masters in France (1886, 70).

Beyond the above-mentioned pedagogical implications of dance, which were also partly taken up and developed in John Locke's *Essay on Education* (1693), Herbert's reflections on dance contain very important semiotic and social considerations. The great diffusion of the practice of dance as a physical exercise in early modern England is justified by the fact that the terpsichorean exercise helps the body to assume an upright posture that the author describes as beautiful to look at. A young gentleman finding himself at Court or any other elitist environment for the first time would take on a more authoritative, graceful, and agile attitude if he were taught to dance properly. Mastering the potential of one's own body, states Herbert, has positive repercussions on one's self-confidence and increases a young person's chances of making his way in society. It was fundamental for a student of law at the Inns of Court to appear confident in society and the practice of dance would guarantee him greater confidence in paraverbal language. After all, even Baldassar Castiglione's *Il Cortegiano* (1528) had recommended dance as an exercise in posture and in awareness of the expressive bodily abilities of the perfect courtier. At the same time, in the dance manuals of Caroso, Arbeau, and Negri, the same hypotheses were advanced about the physical benefits of dancing.

John Playford's *The English Dancing Master* (1651)

Both Italian and French manuals on the one hand, and indigenous sources on the other, contributed to the structure and composition of the first true English dance treatise: John Playford's *The [English]*[19] *Dancing Master*, a book that from its first publication (in 1651) to 1728 was reissued eighteen times, thus "going from a small quarto [...] to a three-volume treatise about English country dances" (Ciambella 2020, 31). As stated in the preface, given the centrality of Playford's book for the study of dance in the seventeenth century, the date of its first edition has been chosen as the *ad quem* point of my analysis on dance in early modern plays.

Not surprisingly, John Playford was a musician, printer, and book seller, revolving around the environment of the Inns of Court: indeed, his printing shop was at the Temple Church, between the Inner and Middle Temple. Although little is known about his life, some of his acquaintances clearly indicate that he was a well-known personality in the seventeenth century music panorama and that he could easily access plays and masques from which he certainly drew inspiration for his treatise. Among his other important friends, such as Henry Purcell and John Blow,[20] there was one Henry Lowes, composer of the music for such Caroline masques as Milton's *Comus* and Carew's *Caelum Britannicum* (both 1634) and Playford's son's godfather. Hence Playford seems to be quite familiar with the tradition of Ben Jonson's and Inigo Jones's masques, also because John Benson, Playford's master in the 1640s, was "involved in a legal dispute over the ownership of what must have been Ben Jonson autographs and foul papers and the consequent beneficial entitlement to publish his work" (Whitlock 1999, 553). Moreover, Whitlock has proved that even Richard Brome's plays influenced Playford's treatise, at least in terms of the list of dances named. Indeed, most of the names of the choreographies for the Country Dances inserted in the first edition of *The English Dancing Master* are taken from lines from Jonson's and Brome's plays (562). As the analysis carried out in Part III will try to demonstrate, however, some of the dances listed and described by Playford are also taken from Heywood's plays.

Therefore, most of the choreographies introduced and explained in *The English Dancing Master* are taken from the Jonsonian tradition of court masques and the plays of Richard Brome. However, despite Playford's intent to popularise and dignify the most indigenous among British dances, his treatise shows "an apparent obsession with French fashion at the English court" (Whitlock 1999, 573), probably due to Queen Henrietta Maria's French influence on the Stuart terpsichorean panorama. For instance, the 49th Country Dance out of the 105 listed in the first edition of the treatise is called *À la mode de France*.

However, what is really striking from a chronological perspective is the publication and immediate success of a dance treatise by a royalist right after King Charles's beheading. In fact, in November 1649, two years before the publication of *The English Dancing Master*, Playford received an arrest warrant – whose execution was never documented – for his publication of *A Perfect Narrative of the Whole Proceedings of the High Court of Justice in the Trial of the King in Westminster Hall*. Nevertheless, it appears somewhat strange that a terpsichorean treatise about the English Country Dances should be published in London while theatres were closed, on the eve of Cromwell's Protectorate in 1653, by a convinced royalist. It is worth noting, moreover, that *The English Dancing Master* went through three editions before the monarchic Restoration in 1660 (1651, 1652 and 1657). To undo the Stuart kings' measures about playing and dancing on Sunday afternoons outlined in the so-called *Book of Sports* (James's first edition in 1617, reissued by Charles in 1633), as will be seen later, Parliament in 1657 passed *An Act for the Better Observation of the Lord's Day*, an act that prevented people from dancing, singing, and playing music on Sunday afternoons, after the mass. Since "private amusement was a matter of private conscience" (Dean-Smith and Nicol 1943, 133) and there were no official restrictions about dancing in private houses and estates, *The English Dancing Master* could have evaded the Puritan censorship thanks to its apparently innocent content, which would have been unlikely to give rise to issues of public order (Whitlock 1999, 553).

Notes

1 In the late 1960s, the Committee on Research in Dance (CORD), publishing the proceedings of the preliminary conference on research in dance, admitted that no academics were yet involved in systematic research on dance at the university level (Bull 1967).
2 Of course the milestone studies on the corporeal turn are those by Maxine Sheets-Johnstone who, in such volumes as *The Phenomenology of Dance* (1966 and ff. editions), *The Primacy of Movement* (1999), and *The Corporeal Turn: An Interdisciplinary Reader* (2009) have made a fundamental contribution to the research on dance and laid the groundwork for a scientific, systematic approach to dance studies.
3 She reaffirmed her claim more recently, specifying that "there is an absence of surviving English dance manuals written between 1500 and 1651 that could compensate for the lack of choreographic specificity within the plays" (2019, 21).
4 George Villiers was one of King James's – subsequently of his son Charles's – favourites and a highly skilled dancer, well-known above all for the elasticity and elegance of his legs, whose musculature, particularly predisposed to dance, was always highlighted in the paintings depicting him (see Williams 2006, 169).
5 If one takes into account Caroso's 1581 first edition of his *Il ballarino*, still considered the first European dance treatise ever published, and de Montagut's and de Lauze's 1620s manuals.
6 The subtitle of *Nobiltà di dame* reads "Libro, altra volta, chiamato *Il ballarino*", meaning "book previously entitled *Il ballarino*".
7 It is no coincidence it is also the name of a dance step: "A leap, or a caper in dancing, especially a cabriole" (*OED* n. 1.1).

8 There is actually another manuscript on the practice of dance in the Elizabethan period, but it is outside the courtly context and so will be analysed in a different section of this study.

9 In particular, they are numbers 1, 5, 6, 7, 8 in the list presented below, since numbers 2 and 3 are anonymous and number 4 was compiled by a personality only indirectly linked to the London schools for lawyers.

10 Contemporary sources inform us that early modern plays were represented both in the royal residences of Greenwich, Whitehall, Hampton Court, and in the noble residences – e.g., the Earl of Pembroke's and Southampton's. Therefore, one cannot rule out the possibility that terpsichorean performances were also staged in the country houses of the English aristocracy, as manuscripts and iconographic sources of the time tend to demonstrate (see Mortimer 2012, 341–351; Wilson and Calore 2005).

11 Early modern English spelling has been modernized here and elsewhere for reasons of legibility. Obsolete grammar forms, however, are retained.

12 For a comprehensive study of dance schools and dance at school, see Eubanks Winkler 2020.

13 For an overview of the Master of the Revels' role, see Dutton 1991.

14 Students at the Inns of Court used to attend lessons with notebooks and take notes about everything they heard and noticed (see Sileo 2016); it is thus plausible that notes about choreographies were written during dance lessons or dancing events at the schools of law.

15 Chute reports that a visitor to one of these dance schools, in describing the grace of a dancer, said: "Wonderfully he leaped, flung and took on" (1949, 89).

16 The text of the regulation reads: "There has late been a great increase in the number of dancing-schools established; these have been conducted by persons unqualified both by their knowledge and their morals, and have been set up in suspect places [...]. The Queen is particularly anxious to suppress those who under the pretence of good exercise entice the young to exercise lewd behaviour" (Chute 1949, 89).

17 It is no coincidence that *Twelfth Night* is the Shakespeare play that mentions the most dances.

18 Actually, the fashion for dances of French origin in England continued for several centuries. In the nineteenth century there were still continental influences (such as the quadrille or the waltz) which arrived in noble and middle-class estates from France (see Ciambella 2013).

19 The adjective 'English' disappeared immediately after the 1651 first edition of the treatise.

20 Both the well-known musician Purcell and the Westminster organist Blow attended Playford's funeral in 1686. Moreover, Purcell composed *Pastoral Elegy on the Death of Mr. John Playford* one year after his friend's death.

2 Salome vs David: The early modern *querelle* on dance between Neoplatonists and Puritans

> While dancing masters sought to legitimatize dance through careful scriptural exegesis, historical exempla, and patriotic fervor, textual archives contain more invective against dance than apologia for it (McCulloch and Shaw 2019, 3).

As McCulloch and Shaw hint at in their introduction to *The Oxford Handbook of Shakespeare and Dance*, dance discourse was a controversial topic in early modern England. On the one hand, dancing masters hailed the beneficial effects of the terpsichorean practice on would-be dancers and amateurs, while on the other hand, some currents of thought condemned dance and dancing.

When reflecting on such a controversial issue, the debate about dance in England between the sixteenth and the seventeenth centuries also offers an important tool for analysing important philosophical, religious, and gender-related implications. The controversy between the Neoplatonic philosophy of Italian origin and indigenous Puritan ethics inexorably gave rise to a religious and moral *querelle* between those intellectuals – not only dancing masters – who tirelessly defended the terpsichorean practice and those who instead condemned it as the most lascivious and corrupt physical exercise, capable of satanically mixing men's and women's bodies.

Yet, thanks to this fierce debate, our knowledge of dance in early modern England has been enriched with nuances linked not only to the mere performance of dance steps or to the simple diffusion of continental dances of French and Italian origin, but also to ideological and cultural implications that broaden the metaphorical and emblematic charge of dance, thus semanticising terpsichorean practice. In other words, the Neoplatonic and Puritan anti-terpsichorean discourses about dance increased in significance when they were inserted within early modern plays in order to underline and corroborate a particular invective against or exaltation of one or more characters of one sex or the other.

The philosophical fulcrum around which all the Neoplatonic discourse in defence of dance developed is the myth of the missing/other half of the androgynous being that Aristophanes recounts in the *Symposium*, the best known of Plato's dialogues. According to the story told by the Greek playwright whom Plato fictionalises in his work, human beings were once circular androgynous creatures with four arms, four legs, two heads, and two sexes. The image immediately recalls the famous drawing of the Vitruvian man by Leonardo Da Vinci – an Italian Renaissance artist, not by chance. Zeus, irritated by the perfection exhibited by these androgynous creatures, decided to split them into two halves, man and woman, thus condemning them to perpetually try to recreate that primary unity opposed by the gods.

Obviously, in the context of Italian fifteenth- and sixteenth-century Catholic Neoplatonism, the Zeus figure is replaced by that of a merciful and benevolent God who endowed human beings with the ability to love each another. Moreover, Plato's conception of the melding of body and spirit as a return to primordial unity becomes the Christian union stretching towards eternity, a union that reproduces on earth the very same reunion man will have with the divine in the afterlife. Therefore, according to Neoplatonic philosophy, the earthly love between man and woman and their spiritual and physical union mirror man's union with God after death.

Hence, according to early modern English intellectuals influenced by this philosophy, there is no better bond than the one created by two dancing bodies. The Italian writers who had embraced the Neoplatonic school of Giovanni Pico della Mirandola, Marsilio Ficino, and Niccolò Cusano had exalted the practice of dance precisely because of the proto-Christian Platonic union of bodies it represented and that recalled the myth from the *Symposium* described above. Among others, Castiglione, in his well-known *Il Libro del Cortegiano*,[1] recommends that a courtier[2] should not only engage in intellectual exercises or those requiring considerable physical efforts, but also in practices that would distract him – such as dance:

> I would have our Courtier sometimes descend to quieter and more tranquil exercises, [...] laugh, jest, banter, frolic and dance, yet in such fashion that he shall always appear genial and discreet, and that everything he may do or say shall be stamped with grace (1901, 31–32).

He continues: "In dancing, a single step, a single movement of the person that is graceful and not forced, soon shows the knowledge of the dancer" (38). The perfect gentleman must not exceed in showing off before everyone, but must exhibit the correct posture and elegance of movement according to his rank:

> There are certain other exercises that can be practised in public and in private, like dancing; and in this I think the Courtier ought to have a care, for when dancing in the presence of many and in a place full of people, it seems to me that he should preserve a certain dignity, albeit tempered with a lithe and airy grace of movement (86–87).

The only purpose of the courtier's graceful exercise of dance must be his beloved woman's courtship: "Who learns to dance and caper gallantly for aught else than to please women?" (220). Therefore, dancing becomes a powerful tool for men's courtship of women. Castiglione's work is a *summa* of Platonism: Plato and his disciple Aristotle are repeatedly mentioned in the book, so much so that both become true pedagogical models for any gentleman's education and good manners. "I am far from sure whether I believe that Aristotle and Plato ever danced", says marquis Gaspare Pallavicino in Castiglione, but "we may believe that they practised what pertains to courtiership, for on occasion they write of it in such fashion that the very masters of the subjects written of by them perceive that they understood the same to the marrow and deepest roots" (286), answers the Genoese doge Ottaviano Fregoso.

In the early sixteenth century in England, Sir Thomas Elyot had already sided with dance practice in his 1531 treatise on moral philosophy *The Book Named the Governor*, dedicated to King Henry VIII. In the first of the three books comprising the volume, Elyot defends the terpsichorean exercise against the attacks of those who "come into the pulpit" (1967, 204), the most intransigent men of faith who would give birth to the Puritan movement in the mid-1500s. According to the English diplomat, not all dances are reprehensible, as the Puritans claimed: after all, even King David danced before the Ark of the Covenant as a tribute of praise and thanksgiving to Yahweh (2 Sam 6.14–16), a biblical episode which is often quoted in anti-Puritan writings in favour of dance.

Elyot's adherence to Neoplatonism is demonstrated by his description of the dance of the celestial spheres, a concept known as *musica universalis*, which actually dates back to Pythagoras and his disciples' philosophy but was embraced by Plato as well:

> The interpreters of Plato do think that the wonderful and incomprehensible order of the celestial bodies, I mean stars and planets, and their motions harmonical, gave to them that intensity, and by the deep search of reason behold their courses, in the sundry diversities of number and time, a form of imitation of a semblable motion, which they called dancing or saltation (1967, 218).

Not only does Elyot discuss the harmonic dance of celestial spheres, but also dancing as an art that mirrors the *musica universalis* and celebrates harmony between man and woman as an emblem of the reunion of the primordial androgynous being:

> In every dance, of a most ancient custom, there danceth to gather a man and a woman, holding each other, by the hand or the arm, which betokeneth concord. Now it behoveth the dancers and also the beholders of them to know all qualities incident to a man, and also, all qualities to a woman likewise appertaining (235–236).

Misogynistic issues as conceived of today seem to be absent from Elyot's conception of dancing, in the sense that both men and women contribute equally to producing *musica universalis* on earth. Nevertheless, each sex should contribute to the creation of this celestial harmony according to its place within society; thus it can be assumed that men and women had different roles in choreographies as in life.

In 1595 Sir John Davies, a member of the Inns of Court, published *Orchestra, or a Poem of Dancing*, probably the most influential work about dance in early modern England prior to Playford's *Dancing Master*. It is a poem of 131 heptastichs for a total of 917 iambic pentameters in rhyme royal (ABABBCC), preceded by a sonnet dedicated to one of the poet's friends, Richard Martin. The main argument of the poem is epic and serious and is introduced as if it were a lost fragment of Homer's *Odyssey*. Antinous, the most audacious of all Penelope's suitors occupying Ulysses's court in Ithaca, tries to persuade her to grant him the honour of a ball during which she would choose him as her bridegroom and future king of the isle. In order to succeed, the witty Antinous delivers a passionate speech about the benefits of dance, imbued with rhetoric and Neoplatonic principles.

Choosing Antinous as his spokesperson, Davies wrote his own apology for dance and published it in the same year of Sidney's *An Apology for Poetry*. Even Davies – like Elyot, whom he probably imitates – affirms that dance is a perfect reproduction on earth of the movements of the celestial spheres; hence it reflects the most natural of all the cosmic movements. Dancing is therefore not only lawful, but right and proper, because if the stars and planets enact harmonious choreographies in their movements, so must men and women. It is clear that, as Tillyard rightly maintains, in a disharmonious and chaotic context such as the court of Ithaca invaded by the Proci, Penelope refuses to dance with Antinous, since their dance could not reproduce the order and harmony of the celestial spheres:

> Penelope refuses to join in something that is mere disorder or misrule, and there follows a *débat* between the two on the subject of dancing, Antinous maintaining that as the universe itself is one great dance comprising many lesser dances we should ourselves join the cosmic harmony (1943, 97).

Dancing is not only a mere exercise that makes a gentleman more elegant, refined, and aware of his place in the society he lives in, but it is also a means of moving his soul towards the values of concord and harmony. The practice of both dance and music was hence connoted also by marked ethical and religious nuances that, as mentioned earlier, date back to the dance performed by David in the temple of Jerusalem. The same episode is quoted again in defence of the terpsichorean practice in the 1607 treatise *Conclusions upon Dances, both of This Age and the Old* by John Lowin, the King's Men's actor and Shakespeare's friend. It is a short religious/moral essay that identifies and analyses the biblical episodes where certain characters show their faith to God through dance. This is exactly the case of King David, to whom Lowin dedicates ample space in his *Conclusions*, seen as an example of a well-received biblical dance in honour of God, who willingly accepts the choreutical offering of his servant and rewards him and the people over whom he reigns with a long period of peace and prosperity.

That same year, James Cleland, in his *Institution of a Young Noble Man*, recommended dance as a good physical exercise, thus entering the long-standing tradition of European conduct books – Castiglione's *Il Cortegiano* is the paramount example – that endorsed dancing as an excellent activity to keep fit. Nevertheless, Cleland deals with many other conceptions related to dance, such as its connections to pagan rites, its relationship with the *musica universalis* and the harmony of the celestial spheres, even the risks connected to the dissolute practice of dance.

The same conclusions were reached by Richard Brathwait(e) in his *The English Gentleman* (1630), where the poet praises the terpsichorean exercise and recommends it to any gentleman wishing to enter the Court and its elitist environment. However, Brathwait's is probably the most blatant case of gender discrimination in the early modern period. Indeed, the following year, 1631, he published *The English Gentlewoman*, which firmly states that women who dance "shall in hell roar terribly and howl miserably" (77) and definitely discourages noblewomen from taking part in dancing events. As Winerock (2019) observes, "unlike Castiglione, who recommends dancing for both male and female courtiers, Brathwaite's approbation is gendered. He condones dancing for gentlemen but condemns it for gentlewomen" (25). However, Brathwait seems to ignore that most of the early modern dances were performed by male-female couples – although same-sex choreographies were allowed – so it was odd to recommend dancing for men while at the same time excluding women from such a practice.

As already mentioned, in addition to dance aficionados who adhered to the Neoplatonic fashion in England, there was also a school of thought that strongly opposed the legitimacy of the terpsichorean practice, in keeping with the precepts of the most extremist branch of English Calvinism: Puritanism. Without dwelling on the rise and development of this religious movement, suffice it to say that Puritan thought was never persecuted in England, nor did Queen Elizabeth openly support it. Indeed, the last of the Tudor monarchs had understood the political and social weight of the Puritans ever since her Elizabethan Religious Settlement of 1559 (reinforced in 1563) was issued. Elizabeth I always tried to exploit Puritan influence by ensuring that they would not come to power (as would happen under Charles I Stuart with Cromwell's revolution).

Drawing on gender-related issues, the English Puritans claimed that dance was a diabolical exercise because women exposed their legs while making turns, twirls, and jumps (especially during such witchy balls as La Volta); moreover, the physical exercise that dance required made a great deal of blood flow to the face, which became redder – an unacceptable scandal that fed male excitement in public and semiotically recalled the sexual libido. Therefore, Puritans countered David's dance before the Ark with Salome's in Matthew 14:3–11 and Mark 6:17–28, where she performs her sensual dance of the seven veils before her excited stepfather Herod, who had granted her the head of

John the Baptist as reward for her performance. David's dance before the Ark of the Covenant was acceptable because it was performed by a single man, not by men and women together, and thus avoided the taint of promiscuity.

Of course, these considerations give rise to gender-related issues, given that anti-terpsichorean intellectuals associated female dancing with Salome, the vicious daughter of an unjust king whom she seduces through a lascivious dance in order to obtain the beheading of a man, indeed of a prophet. On the other hand, not only was David's performance tolerated because he danced to thank God, but also because he was a (right) man. McCulloch and Shaw (2019) talk about the "vertical distinction between the sexes [...] and obedient comprehension of gender roles" (3) to indicate the right place each sex had in early modern dance. Actually, conduct books and dance treatises from the Continent seem to deal in equal measure with gentlemen and noble women, at least from a lexical point of view (see Ciambella 2020, 37–38). This, I would argue, is probably due to the influence of the Neoplatonic ideal of men and women's unconscious nostalgia for the primordial androgynous human being exerted over Renaissance conduct books and dance manuals (for instance Elyot's *Governor* or Davies's *Orchestra*, as seen above). Nevertheless, women are often depicted as static partners who must wait for gentlemen to make the first move, both in the dance routines and in life. Thus Brown and McBride write that "[m]en tended to have more active roles to play in courtly dances than women" (2005, 291). When men bowed, women bowed in response, and when men gave women their arms, they kindly took it in order to begin a choreography, thus activating a monodirectional movement from the man, the motor of the action, to the woman, who had to wait and respond. However, "[f]or wealthy women, dancing was an opportunity to show off their dresses; engage in flirtation; and display their elegance, gracefulness, and knowledge of the latest dances" (Brown and McBride 2005, 291). Therefore, social events like courtly dances highlighted both male and female stereotyped roles: on the one hand, men displayed their good manners and strength, while on the other hand, women exhibited their fashion tastes and grace.

In general, even the most observant Christian writers and philosophers – not necessarily Puritans – judged dancing immoral and used the Holy Scriptures to attack the terpsichorean exercise, especially if carried out by women. This is the case, for example, of the Anglican cleric John Northbrooke, who, in his *Treatise Against Dicing, Dancing, Plays and Interludes, with Other Idle Pastimes* (1577) – considered the first treatise against theatres and dance ever written in England – denounces dance as one of the greatest evils of mankind and corroborates his thesis with quotations and examples from classical authors. Moreover, *The School of Abuse* (1579) by Stephen Gosson, an attack against the "caterpillars of a commonwealth" – as its title page reads – and *The Anatomy of Abuses* (1583) by Philip Stubb(e)s, "portray [dance] as the most pernicious iniquity of the age", according to Brissenden (2001, 13).

The corruption deriving from the promiscuous – rather than harmonious – practice of dance is also the subject of *The Brief Treatise, Concerning the Use and Abuse of Dancing* by the Italian Augustinian monk Pietro Martire Vermigli, translated into English around 1580 by one I. K., who, quoting from the ninth chapter of the *Sirach*,[3] advises a just man against the company of a female dancer to avoid being the victim of vile deceptions. The same conceptions expressed by Peter Martyr in his 1580s treatise are at the origins of the anonymous *A Treatise of Dances Wherein It Is Shewed, That They Are As It Were Accessories and Dependants (or Things Annexed) to Whoredom* (1581), whose title leaves little room for (mis)interpretation.

In his *A Book of Notes and Common Places* (1581), the Calvinist theologian and musician John Merbecke devotes a section to dance, railing against it because it implies that "men should dance mingled together with women" (2003, 284). In the Bible, recalls Merbecke as others before him, Mary, Moses's sister, had not danced with men and King David had not performed in the company of women. St. John Chrysostom had forbidden dancing at weddings because not only did the dance of the bride and groom corrupt the couple itself, but their entire family. However, Merbecke also seems to open a window on the righteousness of dance, stating that with "moderate dancing, we may testify the joy and mirth of the mind" (284) and "by the word Dancing, there is not meant every manner wantonness or ruffianly leaping and frisking: but a sober and holy utterance of gladness" (285).

The same chapter of the *Sirach* that Peter Martyr had quoted to corroborate his anti-terpsichorean invective is cited by the Puritan Christopher Fetherston in his *Dialogue against Light, Lewd, and Lascivious Dancing* (1582), a conversation between a young man and his teacher about the dangerous lasciviousness involved in the practice of dance between men and women – a kind of upside-down *Orchésographie*. Fetherston also finds it impossible for a man to have to dress so ridiculously to dance with a female partner. Clearly, as John Saward (1980, 101–102) observed, Fetherston's invective is aimed at eclipsing the pagan roots of courtly dances, which date back to the ancient propitiatory dances and were considered "in worship of the devil" (Arcangeli 1994, 144).

These same accusations are taken up by another Puritan, Dudley Fenner, who in his *A Short and Profitable Treatise, of Lawful and Unlawful Recreations* (1590) once again offers an example of biblical intertextuality in support of his anti-terpsichorean position, citing Salome's lascivious dance.

To conclude this brief review of anti-terpsichorean – and I would add *ante litteram* misogynistic – thinkers, it is worth mentioning a most influential personality, the Oxford theologian John Rainolds (or Reynolds), one of the initiators of the *Authorized Version* of the Bible in 1611. In 1599, Rainolds published his *Overthrow of Stage-Plays*, where, like many Puritans before him, he called into question the episode of Herodias performing the dance of the seven veils for Herod. Rainolds's exegetical mistake is serious and obvious, since it is Salome, Herodias's daughter, who performs that dance, not her mother. This is a mistake which, according to Brissenden, "indicates the lack of Biblical example on which the anti-dance critics could draw" (2001, 13), even if in this case it would seem to be rather a mere error in reading the sacred text or a very banal confusion on the part of the author who, being one of the most eminent Puritans of his day, certainly possessed a thorough knowledge of the Scriptures.

It was a tradition in Medieval and Renaissance Europe that "after the sermon and the sacraments in the morning, villagers lounged or played on Sunday afternoon. [...] Wine or ale, music, and dance accompanied the peasant games and frolic" (Baker 1988, 45). Nevertheless, Puritans were ferociously against such impure exercises on the Lord's day. In 1617, Thomas Morton, bishop of Chester, asked King James to intervene in a long-standing dispute between Puritans and some Roman Catholic noblemen about Sunday sports. King James, acknowledged enemy of the English Puritans, answered by issuing *The Book of Sports* (or *Declaration of Sports*), which decreed that dancing was among the permissible sports on Sunday afternoon.

In 1633, as Puritans were gaining power in Parliament, King Charles I reissued his father's *Book of Sports* as an anti-Puritan act of force and renamed it *The King's Majesty's Declaration to His Subjects Concerning Lawful Sports to Be Used.* Nevertheless, the book was publicly burnt in 1643, in the middle of the Civil War.

Notes

1 The book was translated into English by Sir Thomas Hoby in 1561.
2 Although Castiglione addressed his book primarily to male Italian courtiers, he considered dancing a good physical exercise both for men and women.
3 "Use not much the company of her that is a dancer, and hearken not to her, lest thou perish by the force of her charms" (*Sirach* 9.4).

3 Folk and courtly dances

Most of the dances named in the sections above fall under courtly dances, choreographies performed at Court or in the aristocratic estates during festivities or theatrical events. Although Queen Elizabeth used to entertain herself with folk dances, and despite her Stuart successors' love of disguising themselves and their entourage as peasants and shepherds during masques, the distinction between elitist and popular dances remained firm in early modern England.[1] Respect for the etiquette and for conventions established in dance manuals was needed at Court and in the Inns of Court, while the spirit of English folk dances was much freer and more spontaneous, although paradoxically more subject to the influence of the moral dictates of Puritanism. According to Julia Sutton, "the line dividing court and folk dance appears to have been simply a separation marked by greater attention to elegance, style, and technique on the part of those of gentle birth" (1995, 29).

Therefore, not only is the comparison between folk and court dances open to social interpretations, but to moralistic and religious ones as well, especially in the English countryside, where daily life revolved around parishes. In such environments, ancient pagan rites of Celtic and Roman origin, Christian dances, and Puritan restrictions blend together, giving rise to a terpsichorean panorama still neglected by critics, as Winerock rightly complains (2005, 36; 2019). She examines a fair number of parish registers of the English countryside – what she calls "aggregated archival records" (2019, 30) – that testify to the performance of some folk dances that never entered the Court. Thanks to these records, many English folk dances have been discovered that otherwise would not have come down to us, such as the Long Dance, performed during religious ceremonies by villagers and their mayors, or the Furry Dance, a propitiatory dance performed by a long line of people who danced by houses and gardens to purify villages from evil spirits. In the aggregated archival records we also find the Cushion Dance, a promiscuous choreography in which men and women danced together and even chose their partner with a kiss, as reported in the *Gloucester Diocese Consistory Court Deposition Books*.

Another "immoral" dance, performed by one Henry Pillchorne with his trousers down to his ankles, is referred to as the Piddecocke in the *Ex Officio Act Book* of Bridgwater, Somerset. As Winerock points out, this is likely wordplay for the Wooddicock, a dance also mentioned by Playford. Other dances of popular origin are the Lewd Dance, Cross-dressed Dance, and Rope Dance, whose names indicate the kind of dance and the tools or accessories used to perform it.

One the most popular dance of folk origin performed at the Tudor and Stuart Courts is the Jig (Vuillier 2004, 85), a choreography with very fast steps and pressing rhythms that was inserted at the end of plays and in most cases performed by the clown. Indeed, in 1599, the Swiss traveller Thomas Platter may have witnessed a Jig after attending a performance of Shakespeare's *Julius Caesar*:

> On September 21st after lunch, about two o'clock, I and my party crossed the water, and there in the house with the thatched roof witnessed an excellent performance of the tragedy of the first Emperor Julius Caesar with a cast of some fifteen people; when the play was over, they danced very marvellously and gracefully together as is their wont, two dressed as men and two as women (cit. in Williams 1937, 166).

Although whether it was in fact a Jig is unclear, it is well known that this dance was performed at the end of early modern plays. This is also supported by the fact that in 1612 the Middlesex magistrates forbade "lewd jigs" (Smith 1999, 58) at the end of theatrical performances at the Fortune, Curtain, and Red Bull theatres.

However, the dance that more than any other came to symbolise the British indigenous folk tradition is undoubtedly the Morris, a ball still performed today in British folk festivals. The Morris Dance also became famous in the sixteenth century thanks to William Kempe's autobiographical pamphlet *Kemp's Nine Days Wonder. Performed in a Dance from London to Norwick* (1600). The famous actor of the Chamberlain's Men (until 1599), one of the best performers of Shakespeare's clowns, performed a unique Morris Dance marathon from London to Norwich in nine days in 1600 (as the title of his work reads) simply to prove to Queen Elizabeth that he had been wrongly expelled from the Chamberlain's Men and that he was still at the height of his powers and ready for new employment. According to Kempe, a large crowd joined his performance, following him to the town of Norwich, and this is enough to underline the popular character of the Morris Dance. Although Winerock states that Kempe himself, far from belonging to rural environments, was a man associated with Elizabeth's Court (2005, 36), nevertheless his Morris marathon is certainly linked to English folklore, since it was performed outside the Court, on the road from London to Norwich, and members of the lower classes accompanied him. The popular character of the dance in question has been vouched for on more than one occasion since the fifteenth century (see Lancashire 2002, 278; Heaney 2004) and the name Morris has continued to be associated with British tradition and folklore until the present day, with the establishment and flourishing of cultural associations such as The Morris Ring and The Morris Federation, which cultivate an interest in the authenticity of this traditional folk dance.

Another important folk dance in Medieval and early modern England was the Maypole, whose origins date back to ancient Greece. Given its origins, this ancient folk dance was thought to express pagan values, judging by the fact that in 1664, even after the monarchic Restoration, the Puritans managed to ban it throughout the British Isles (O'Connor 2009, 130). In the English Middle Ages, villagers would rise at dawn on the first day of May to go and cut down a pine tree that would become the pole around which they would perform their propitiatory dance. The Maypole was very popular in England: for instance, Thomas Hardy's *Tess of the D'Urbervilles* (1891) begins with a Maypole.

However, if the majority of English folk dances were banned by the Puritans or were the target of their stinging criticism, a different fate befell courtly dances, which evolved "into a means of courtly self-fashioning" (Howard 1998, 3), in part because of the support and approval of three monarchs who liked to be entertained by performances at Court and who even tried their hand at dancing, as will be shown below. Neoplatonic ideas – which entailed a certain cultural awareness and at least a discreet competence in the Italian language – did not succeed in permeating the lower strata and remained confined to the cultural elites of the universities and the Court. Courtly dances thus became the tangible manifestation of Plato's world of ideas, an art capable of ennobling human beings to the point of raising them to heaven, to the ethereal world of ideas.

Even at a semiotic level, the human body assumes the necessary posture to become an emblem of ideal Neoplatonic perfection. Gentlemen must be erect and confident, with a proud gaze and the face slightly inclined upwards, often in the gallant act of extending their arm or hand to a female partner. Everything in their figure leans upwards, towards the sky, towards that world of ideas that is as aloft as the celestial spheres (see Howard 1998, 1–2). While folk dances, at the same time chaotic and spontaneous celebrations by a social class that owes much to the land, are downward dances – far from elegant choreographies – performed to bring about the success of the harvest, courtly balls thank a benevolent deity who has allowed the sovereign and his closest collaborators to reign and enjoy the power conferred on them from above.

The discomposure and the lack of rigid and fixed schemes typical of folk dances are therefore opposed to the harmonious choreographies of the court dances, such as the Old Measures mentioned earlier. However, upper-class choreographic performances are not synonymous with slow, cadenced rhythms, as one might think. It is true that, on the one hand, such dances as the Pavan or Saraband required slow and elegant choreographies suitable for solemn ceremonies such as weddings, on the other hand, however, it was customary to entertain the Court even with the tightest rhythms of the Galliard or La Volta.

Moreover, a cultural historical element separates the terpsichorean performances of peasants from those of the courtiers: the origin of the various dances. Popular dances are in fact the oldest and undoubtedly the only truly indigenous English ones. Indeed, peasants and villagers could certainly never enjoy close connections with the continental courts. Therefore, if the origins of the Morris or Maypole dances can be traced to a common European past,[2] other folk dances have their roots in British folklore, as in the case of propitiatory dances performed before, during, or after the harvest.

Conversely, in the early modern English Court, fashionable dances were mainly from France and Italy. Aside from a few musical and/or step variations, the English Pavans and Galliards are in no way of indigenous origin. The English Court was subject to continental fashions – to the disappointment of some prominent personalities, as seen above. This cultural and historical issue is part of a lengthy campaign promoted by intellectuals in the spirit of British nationalism which began with Henry VIII and his schism from the Roman Catholicism, passing through James VI and I and his union of the two crowns in 1603, and culminating in the following century with the unification of the parliaments of Scotland and England in 1707 (Macinnes 2007).

The only court dance partly of English origin is the Country Dance, born from the fusion of popular rhythms and choreographic positions imported from Europe, thanks also to the intervention of Queen

Elizabeth who loved to be entertained with folk performances at Court (cf. Nichols 1823). It is no coincidence that when the British nationalistic spirit developed further in the seventeenth century, leading to a process of national unification and simultaneous isolation from the rest of Europe, the Country Dance became the terpsichorean emblem of British patriotism,[3] reproducing the British social macrocosm hierarchised in the microcosm of the noble and bourgeois dance floors (cf. Ciambella 2013, 55–63).

The only place where courtly and folk dances converge – except for those occasions when British monarchs summoned groups of popular artists to their Court – is the stage. The heterogeneity of the subjects dealt with by the early modern theatre corresponds both to a variety of characters on stage from all social strata and to a varied audience. In fact, on the one hand, upper classes could identify with the masked ball in *Romeo and Juliet* (1.5) – probably a Pavan or certainly one of the other Old Measures – while on the other, peasants and craftsmen would be enthusiastic about the frolicsome dance of the twelve satyrs in *The Winter's Tale*, two opposing examples of terpsichorean performances.

Whether the dance was charged with ironic connotations or acted as a quick intermezzo to facilitate scene changes, early modern theatre brought on stage a multiplicity of choreographic performances even within the same play. Consider for instance the folkloristic Bergomask in *A Midsummer Night's Dream*, compared to the solemn and elegant Carols of Oberon and Titania's fairies. The early modern theatrical microcosm reproduced the social macrocosm from every point of view, not only in its political, ethical, or religious intermingling, but also at the terpsichorean level. Dance thus becomes the emblem of a strongly hierarchical society, where the grace and elegance of the rich and powerful Court are contrasted with the liveliness of the chaotic working class.

Notes

1 Nevertheless, court and folk dances had probably been influencing each other since earlier times, although "the degree to which folk and court dance influenced each other during the Renaissance is unknown" (Sutton 1995, 29).

2 Examples of these dances can also be found in the Basque Country or Spain, probably due to the presence of the Celts in Western Europe or the effects of the Roman conquest, rather than direct influence.

3 Suffice it to note that Playford's *The English Dancing Master*, the first collection of English Country Dances, went through eighteen editions between 1651 and 1728, as stated elsewhere.

4 From Elizabeth to Charles, through James: Dance and politics

> As an important cultural practice, dancing reflected and participated in the broad social and political changes of early modern England: the rise of the centralized state, the emergence of the patriarchal family, the polarization of religious factions, and the acceleration of exploration and colonization (Howard 1998, 3).

Skiles Howard perfectly summarises all the issues at stake surrounding the complex relationship between events and political changes and the practice of dance in sixteenth- and seventeenth-century England. Dancing embraces every area of the early modern English cultural panorama, from religious to social and political spheres. The relationship between dancing and politics, mainly of a propagandistic nature, is of paramount importance given that three sovereigns – first Elizabeth Tudor and then James and Charles Stuart – are personally involved in the practice of choreographic exercise, both as real physical training and as their favourite form of entertainment.

The energetic Queen Elizabeth I loved to try her hand at dancing as a physical exercise. The most famous and pertinent statement in this regard is undoubtedly the testimony given by Sir John Stanhope, gentleman of the Queen's Privy Chamber, who, writing to Lord Gilbert Talbot on 22 December 1589, informs his friend about Elizabeth's physical fitness: "My Lord, the Queen is so well as I assure you, six or seven galliards in a morning, besides music and singing, is her ordinary exercise" (cit. in Lodge 1838, 386). Galliards required considerable physical preparation, given the liveliness of their rhythm and their quick sequence of steps. Moreover, the famous painting entitled *Queen Elizabeth I Dancing with Robert Dudley, Earl of Leicester*, kept at Penshurst Place in Tonbridge, Kent, appears to portray Her Majesty performing La Volta with one of her favourite courtiers.[1]

Thus, while the Tudor monarch assiduously exercised with complicated and dynamic choreographies, she did not disdain to attend terpsichorean events, especially when organised by Robert Dudley, a true connoisseur of dance and an excellent organiser of events at Court (Frye 1996, 66–72), or by another dancer she deeply admired, Sir Christopher Hutton, who "was taken notice of by [her] for his gracefulness in dancing before her at court, which proved the first step for his future preferments" (Nichols 1828, 185).

Elizabeth also had her own entourage of artists who entertained her with performances at Court and on tour, the Queen's Men, a company of professionals who in recent years have been enjoying increased attention by scholars of Elizabethan theatre (see McMillan and MacLean 2006; Ostovich et al. 2016). Founded in 1583 on the monarch's initiative, the Queen's Men company comprised actors taken from other existing stable companies, such as the Dudley's, the Leicester's, or the Sussex's Men. In 1584 the company asked the Queen's Privy Council for permission to perform in public, which was not granted (Montrose 1996, 198). The company then continued to perform at Court during the Christmas holidays, or in the main English cities and provinces, until 1603, when it was dissolved by James I.

Elizabeth's propensity for dancing extended beyond courtly performances. In this regard, it is important to mention Nichols's report of the Queen's good disposition towards performances of folkloristic character:

> Her Majesty that Saturday night was lodged again in the Castle of Warwick [...], where it pleased her to have the country people, resorting to see her, dance in the court of the Castell, her Majesty beholding them out of the chamber window, which thing, as it pleased well the country people, so it seemed her Majesty was much delighted, and made very merry (1828, 318).

Her Majesty's passion for non-courtly dances probably led to the spread of the English Country Dance as it was known and danced in the following centuries: a mixture of court and popular terpsichorean paradigms that, as stated above, became the emblem of British nationalism in the long seventeenth century.

However, the shrewd Tudor Queen did not fail to deal with economic issues related to dance and especially to private schools where this art was taught, as evidenced by the aforementioned monopoly regulation limiting the number of non-public dance schools. The importance that dance was acquiring during the years of Elizabeth's reign clearly led her to want to personally control the businesses that revolved around it in the certainty that it would be an assured source of income for the monarchic coffers.

James, like his son Charles after him, was particularly fond of court performances and was a lover of theatre and dance. His reign (1603–25), like his son's (1625–42), is the era of the masque, an ancient type of court performance that under his rule took on new vigour and was enriched by new scenographic elements, mostly thanks to the eighteen iconological masques by Ben Jonson, in collaboration with the famous royal designer Inigo Jones.[2] According to Jerzy Limon (1990, 123), the Jonsonian masque celebrated James's "divine wisdom" by representing him as *Rex Pacificus*, a peaceful and pacific king blessed by God, who reigned in splendour and harmony.

The masque was the preferred form not only of the King, but also and above all of the Queen Consort Anne of Denmark and their first-born son, Prince Henry, an excellent dancer who died at eighteen, thus allowing Charles to ascend the throne after his father's death in 1625. The Queen personally ventured into the performances together with her husband, exactly as King Charles and the Queen Consort Henrietta Maria did after their predecessors. In fact, Queen Anne's name is often included in the dramatis personae of the masque, as in the case of Jonson's *Masque of Blackness* (1605), where Anne played Euphoris, one of the twelve messengers of the sea, or tritons. In the court masques, dance was central and, as is well known, involved not only the actors on stage but also the audience of gentlemen and noblewomen, who were invited to take part in the events interpreted by spirits, nymphs, or other ethereal creatures.

James's and Anna's love of dancing is also reflected in the staging of Jacobean plays. It is no coincidence that, from 1603 until his last solo play, *The Tempest*, Shakespeare included dance scenes in any play he wrote and that was staged at Court, also proposing masques – as for instance in the case of *The Tempest* – within the dramatic plot.

When he ascended the throne in 1625, Charles I continued his father's politics of spectacularisation of courtly performances, so much so that when referring to Jacobean and Caroline masques, critics generally use the wide-ranging label "Stuart masque" (Bevington and Holbrook 1998). However, unlike the original Jacobean masque, the Caroline one was thematically enriched by a precise paradigmatic and polemical intention against Charles (Guardamagna and Anzi 2002, 143) and the hyper-centralised kingdom of the *Rex Bellicosus*, as one sect of his enemies – the Independents – used to call him. Moreover, from a structural viewpoint, new French and Italian tastes and dances were brought to the Court of London by Queen Henrietta Maria, daughter to Henry IV of France and Maria de' Medici. This raised the staging costs of masques at Whitehall to some £21,000, with a cast of almost 1,000 actors, as in the case of Shirley's *The Triumph of Peace* (1634), thus intensifying the people's dissatisfaction with Stuart politics. In the Caroline era, therefore, dance became a magnificent form of spectacle at Court and choreographies were danced by many more dancers than previously, thus contributing to the increased complexity of the choreographies themselves. Of course, with the Puritans' ascension to the English Parliament and the closure of theatres in 1642, even the pompous Stuart masques ceased to be staged at Whitehall. The last masque produced at Court was William Davenant's *Salmacida Spolia*, designed by Inigo Jones and performed at Whitehall on 21 January 1640. After that date, as the Parliamentary newsletter *Mercurius Rusticus* observed in October 1643, "the Queen will not have so many Masques this Christmas and Shrovetide this year as she was wont to have other years heretofore; because *Inigo Jones* cannot conveniently make such Heavens and Paradise at Oxford as he did at *White-hall*" (cit. in Hotson 1928, 9; emphasis in the original). Although Jones had already taken his machineries to Oxford in 1636, where Charles and Henrietta Maria would move the royal capital after 1642, the pomposity of the courtly masques at Whitehall was definitely over, despite some musical and dance performances held in Oxford as "clandestine revivals" (Stubbs 2011, 356).

Therefore, on the one hand, for Elizabeth Tudor dance had become a form of political propaganda, whereas James and Charles Stuart used it as an authentic celebration of royalty by divine will, given the increasing preoccupation with and discontent for their political inability. The harmonious courtly dances that reproduced the heavenly paradigms in Elizabethan England became the sign of the (albeit false) perfection of James's and Charles's Court, a Court blessed by God himself.

Like Elizabeth, even James and Charles were not spared the vitriol of critics who saw in this propensity for dancing and peace a sign of weakness. To their accusers, the Stuart Kings' love of dancing was a symptom, on the one hand, of their homosexuality, which gave their inept favourites political positions of considerable prestige. On the other hand, performances at Court contributed heavily to the public spending of a Court that was falling headlong into blatant corruption and waste, which would be condemned more or less directly in many writings of the time (Guardamagna and Anzi 2002, 12). Moreover, James's predilection for such frivolous entertainments as dancing and for same-sex relations – especially with George Villiers, first Duke of Buckingham and an acknowledged skilful dancer – resulted in accusations of homosexuality, something that led his son Charles to dislike Buckingham during the first years of his reign.[3]

Thus we have seen how early modern dance becomes emblematic of a harmony and order constantly sought after in every sphere and at every cultural level. This propensity is demonstrated by all the written productions of this period, from Spenser's hymns to Shakespeare's *Troilus and Cressida*, from religious homilies to Sir Walter Raleigh's pompous *History of the World* (Tillyard 1943, 7–15). After all, at this time England is deeply divided and marked by the religious questions posed by the Reformation, by events and wars, and lastly by the discovery of a cosmological system that no longer places the Earth and humankind at the centre of the universe, but that downgrades our planet to a simple celestial body sharing with others a revolution around a central Sun. Ultimately, the perfection and rigour of the positions of court dances represented a paradigm of order and harmony at both a social and individual level, of music of the celestial spheres reproduced on Earth, as early modern texts – from Elyot's *Governor* and Davies's *Orchestra*, to Jonson's masques, Milton's *Arcades* (1634)[4] and, ultimately, even Davenant's *Salmacida Spolia*[5] – suggest.

Such ordered and harmonious paradigms should also be reflected at a textual linguistic level, as the next chapters will demonstrate, in early modern plays, where dance discourse was highly emblematic of harmonious or disharmonious relationships among characters, genders, and social classes, thus exhibiting "the range and variety of contemporary conceptions of dancing" and "dancing's multiple meanings and functions" (Winerock 2019, 25).

Notes

1 Actually, modern critics deny this interpretation and believe the painting does not portray the Queen, but merely an ironic representation of the English way of dancing, since the anonymous painter might be associated with the French Valois school (see Holman 1983, 53).

2 While Jonson died in 1637 and so could not have worked in the last years of Charles's reign, Inigo Jones, who died 1652, designed court masques until the end of the Caroline period and beyond.

3 For a detailed description of Buckingham's role as masquer and masque commissioner, and his relationship with James I and Charles I, see MacIntyre 1998.

4 Cf., for instance, vv. 62–67: "listen I / To the celestial *Sirens* harmony, / That sit upon the nine enfolded Sphears / And sing to those that hold the vital shears / And turn the Adamantine spindle round, / On which the fate of gods and men is wound".

5 Cf., for instance, the first stanza of the 6th song: "To the King and Queen, by a chorus of all. / So musical as to all ears, / Doth seem the music of the spheres, / Are you, unto each other still; / Tuning your thoughts to eithers will" (vv. 1–5).

References

Arbeau, T. (1589) *Orchésographie.* [Online]. Jehan des Preyz: Langres. Available at https://www.graner.net/nicolas/arbeau/ [Accessed on 23 March 2020].

Arcangeli, A. (1994) Dance under trial: The moral debate 1200–1600. *Dance Research: The Journal of the Society for Dance Research* 12(2), 127–155.

Baker, W.J. (1988) *Sports in the Western World. Revised Edition.* Urbana/Chicago, IL: University of Illinois Press.

Bevington, D. and Holbrook, P. (eds.) (1998) *The Politics of Stuart Court Masque.* Cambridge: Cambridge University Press.

Brathwaite, R. (1631) *The English Gentlewoman.* London: B. Alsop and T. Fawcet.

Brissenden, A. (2001) *Shakespeare and the Dance.* Second edition. Atlantic Highlands, NJ: Humanities Press.

Brown, M.L. and Mcbride, K.B. (2005) *Women's Roles in the Renaissance.* Westport/London: Greenwood Press.

Buck, G. (1605) Of orchestice, or the art of dancing. In Stow, J. (ed.) *The Annales, or Generall Chronicle of England.* [Online]. London: Edmond Howes. Available at http://winerock.altervista.org/sources/buck_orchestice.html [Accessed on 23 March 2020].

Bull, R. (ed.) (1967) CORD Dance Research Annual. Research in Dance: Problems and Possibilities. *The Proceedings of the Preliminary Conference on Research in Dance.* New York: CORD Inc.

Chute, M. (1949) *Shakespeare of London.* London: Dutton.

Ciambella, F. (2013) *Testo, danza e corpo nell'Ottocento inglese.* Roma: Aracne.

Ciambella, F. (2017) *"There was a star danced": Danza e rivoluzione copernicana in Shakespeare.* Roma: Carocci.

Ciambella, F. (2020) The [Italian] dancing master: English reception of Italian Renaissance terpsichorean manuals. A corpus-driven analysis. In: Magazzù, G., Rossi, V. and Sileo, A. (eds.) *Reception Studies and Adaptation: A Focus of Italy.* Newcastle upon Tyne: Cambridge Scholars Publishing, 28–44.

Cunningham, J.P. (1965) *Dancing in the Inns of Court.* London: Jordan & Sons.

Daye, A. and Thorp, J. (2018) English Measures Old and New: Dulwich College MS. XCIV/f.28. *Historical Dance* 4(3), 27–40.

Dean-Smith, M. and Nicol, E.J. (1943) *The Dancing Master*: 1651–1728. *Journal of the English Folk Dance and Song Society* 4(4), 131–145.
Dutton, R. (1991) *Mastering the Revels: The Regulation and Censorship of English Renaissance Drama*. Iowa City, IA: University of Iowa Press.
Elton, W.R. (2000) *Shakespeare's* Troilus and Cressida *and the Inns of Court Revels*. Aldershot: Ashgate.
Elyot, T. (1967) *The Boke Named the Governour* (1531). Arlington: Franklin.
Eukbanks Winkler, A. (2020) *Music, Dance, and Drama in Early Modern English Schools*. Cambridge: Cambridge University Press.
Frye, S. (1996) *Elizabeth I: The Competition for Representation*. Oxford/New York: Oxford University Press.
Goose, N. (2005) Xenophobia in Elizabethan and early Stuart England. In: Goose, N. and Luu, L. (eds.) *Immigrants in Tudor and Early Stuart England*. Brighton: Sussex Academic Press, 110–135.
Guardamagna, D. and Anzi, A. (2002) *Storia del teatro inglese. Il teatro giacomiano e carolino*. Roma: Carocci.
Heaney, M. (2004) The earliest reference to the Morris Dance? *Folklore Music* 8(4), 513–515.
Herbert, E. (1886) *The Autobiography of Edward, Lord Herbert of Cherbury* (1643), ed. by S. Lee. Second edition. London: Routledge.
Hoby, T. (1901) *The Book of the Courtier by Count Baldasar Castiglione* (1561). New York: Charles Scribner's Sons.
Holman, P. (1983) The English royal violin consort in the sixteenth century. *Proceedings of the Royal Musical Association* 109, 39–59.
Hood Philips, O. (2005) *Shakespeare and the Lawyers*. London: Routledge.
Hotson, L. (1928) *The Commonwealth and Restoration Stage*. Cambridge, MA: Harvard University Press.
Howard, S. (1998) *The Politics of Courtly Dancing in Early Modern England*. Amherst, MA: University of Massachusetts Press.
Kaeppler, A. (2000) Dance ethnology and the anthropology of dance. *Dance Research Journal* 32(1), 116–125.
Lancashire, A. (2002) *London Civic Theatre*. Cambridge: Cambridge University Press.
Lee, S. (ed.) (1886) *The Autobiography of Edward, Lord Herbert of Cherbury: With Introduction, Notes, Appendices, and a Continuation of the Life*. London: John C. Nimmo.
Limon, J. (1990) *The Masque of the Stuart Culture*. Newark, DE: University of Delaware Press.
Lodge, E. (1838) *Illustrations of British History, Biography and Manners*, Vol. 2. London: John Chidley.
Macinnes, A.I. (2007) *Union and Empire: The Making of the United Kingdom in 1707*. Cambridge: Cambridge University Press.
Macintyre, J. (1998) Buckingham the masquer. *Renaissance and Reformation* 22(3), 59–81.

Mcculloch, L. and Shaw, B. (2019) Introduction. In: Mcculloch, L. and Shaw, B. (eds.) *The Oxford Handbook of Shakespeare and Dance*. Oxford/New York: Oxford University Press, 1–10.
Mcmanus, C. (2002) *Women on the Renaissance Stage. Anne of Denmark and Female Masquing in the Stuart Court 1590–1619*. Manchester: Manchester University Press.
Mcmillan, S. and Maclean, S. (2006) *The Queen's Men and Their Plays*. Second edition. Cambridge: Cambridge University Press.
Merbecke, J. (2003) *A Booke of Notes and Common Places* (1581). Ann Arbor, MI: University of Michigan Press.
Montrose, L.A. (1996) *The Purpose of Playing: Shakespeare and the Cultural Politics*. Chicago, IL: University of Chicago Press.
Mortimer, I. (2002) *The Time Traveller's Guide to Elizabethan England*. London: Random House.
Nichols, J. (1823) *The Progresses and Public Processions of Queen Elizabeth*, Vol. 1. London: John Nichols.
Nichols, J. (1828) *The Progresses, Processions, and Magnificent Festivities of Queen Elizabeth*, Vol. 1. London: John Nichols.
O'connor, S. (2009) *Dance of Language*. Bloomington, IN: Author House.
Ostovich, H., Schott Syme, H. and Griffin, A. (eds.) (2016) *Locating the Queen's Men 1583–1603: Material Practices and Conditions of Playing*. London/New York: Routledge.
Saward, J. (2000) *Perfect Fools: Folly for Christ's Sake in Catholic and Orthodox Spirituality*. Fourth edition. Oxford: Oxford University Press.
Sheets-Johnstone, M. (1966) *The Phenomenology of Dance*. Madison, WI: University of Wisconsin Press.
Sheets-Johnstone, M. (1999) *The Primacy of Movement*. Amsterdam/Philadelphia, PA: John Benjamins.
Sheets-Johnstone, M. (2009) *The Corporeal Turn: An Interdisciplinary Reader*. Exeter/Charlottesville, VA: Imprint Academic.
Sileo, A. (ed.) (2016) *Il diario di John Manningham del Middle Temple*. Roma: Universitalia.
Smith, B.R. (1999) *The Acoustic World of Early Modern England*. Chicago, IL: University of Chicago Press.
Stokes, J. and Brainard, I. (1992) The olde measures in the west country: John Willoughby's manuscript, *Records of Early English Drama* 17(2), 1–10.
Stubbs, J. (2001) *Reprobates: The Cavaliers of the English Civil War*. New York/London: Norton.
Sutton, J. (1995) Late-Renaissance dance. In: Sutton, J. (ed.) *Courtly Dance of the Renaissance*. New York: Dover Publications, 21–31.
Tillyard, E.M.W. (1943) *The Elizabethan World Picture*. London: Chatto & Windus.
Vuillier, G. (2004) *History of Dancing from the Earliest Ages to Our Own Times*. Whitefish, MT: Kessinger.

Wagner, A. (1997) *Adversaries of Dance: From the Puritans to the Present.* Urbana, IL: University of Illinois Press.

Whitlock, K. (1999) John Playford's *The English Dancing Master* 1650/51 as cultural politics. *Folk Music Journal* 7(5), 548–578.

Williams, C. (1937) *Thomas Platter's Travels in England.* London: J. Cape.

Williams, R.P. (2006) *Sweet Swan of Avon: Did a Woman Write Shakespeare?* Berkeley, CA: Wilton Circle Press.

Wilson, C.R., and Calore, M. (2005) *Music in Shakespeare: A Dictionary.* London/New York: Continuum.

Wilson, D.R. (1986–7) Dancing in the Inns of Court. *Historical Dance* 2(5), 3–16.

Winerock, E.F. (2005) Dance references in the records of early English drama: Alternative sources for non-courtly dancing, 1500-1650. In: Cook, S.C. (ed.) *Proceedings of the 26th Society of Dance History Scholars Annual Conference (June 17–19, 2004).* Riverside, CA: Society of Dance History Scholars, 36–41.

Winerock, E.F. (2011) Staging dance in English Renaissance drama. In: Pierce, K. (ed.) *Proceedings of the 34th Society of Dance History Scholars Annual Conference (June 23–26, 2011).* Riverside, CA: Society of Dance History Scholars, 259–266.

Winerock, E.F. (2019) "The heaven's true figure" or an "introit to all kind of lewedness"? Competing conceptions of dancing in Shakespeare's England. In: Mcculloch, L. and Shaw, B. (eds.) *The Oxford Handbook of Shakespeare and Dance.* Oxford/New York: Oxford University Press, 21–48.

Part II

Dance and/as language: State of the art and methodological issues

As noted in the preface, this second part focuses on dance discourse – in a Foucauldian sense – and thus hypothesises the existence of a language *for* dance, i.e., reading dance as language, and a language *about* dance. In other words, when dealing with the relationship between dance and language, I would argue, at least two different approaches have been used by critics. On the one hand, some scholars treat dance as a (bodily) language, thus seeking a language *for* dance, and establish a series of rules that apparently connect the movement of the dancing body to those governing any natural language. The studies belonging to this branch, mainly covered by dance semioticians, theorists, and choreographers, share the idea that a so-called grammar of dance exists, which has indisputable similarities with language grammar rules. On the other hand, both corpus and textual linguists tend to use the methodological tools offered by such fields as critical discourse analysis, stylistics, and corpus linguistics to analyse morphosyntactic, lexicosemantic, and the pragmatic aspects of dance scenes as they emerge in novels, poems, plays, and other textual genres. The primary objective of these analyses is often to identify possible recurring patterns or, by drawing on literary theories as well – as in the case of dance lexis presented here – co-textual elements that enable one to understand even the metaphorical, allegorical, symbolic, and emblematic charge of dance scenes. In the case of the early modern English period under scrutiny here, I would say that mainly thanks to the ideas put forward by Neoplatonic philosophy and resulting from Puritanical attacks, dance is a polysemic art that includes a multiplicity of meanings and functions in its paraverbal language, from emblem of harmony between men and women to a mirror of the movement of the celestial spheres – or an exercise that corrupts both mind and body. This polysemy will be investigated in the lexical analysis carried out in the next part of this book.

5 A language *for* dance: Dance as language

Probably the first study to treat dance as an out-and-out language with its own rules to be followed is *Grammatik der Tanzkunst*, an 1887 treatise by the nineteenth-century German dancer, choreographer, and theorist Friedrich A. Zorn. His book was translated into English in 1905 by the American dancing master Alfonso J. Sheafe as *Grammar of the Art of Dancing*. Zorn's study focuses on both music scores and steps for different choreographies, thus combining notes on the stave with positions of the arms, legs, and feet. As the author himself declared in his preface (1905, xii), exactly like the music of Mozart, Beethoven, or Wagner, even dance needs a script to be followed and that could be archived. That is the idea behind a grammar of dance whose rules might be defined and observed. Zorn's attempt to establish steps for various choreographies might be compared to Italian Renaissance manuals, English manuscripts describing the eight Old Measures at the Inns of Court, or Playford's schematisation of country dances discussed above. Nevertheless, contrary to the sixteenth- and seventeenth-century directions about dancing, which were addressed to gentlemen and gentlewomen who needed to learn how to dance, *Grammar of the Art of Dancing* is a scholarly manual addressed to experts, not to the "average reader" (xi).

The idea of dance as language continued to be cultivated during the early twentieth century, when the greatest Hungarian-born dance theorist, Rudolf Laban, invented a proper language *for* dance, which was named Labanotation (or kinetography) after him.[1] In 1928, Laban published his notation in the first issue of the journal *Schrifttanz*, which he funded in Germany and which was published until 1931. Without going into too much detail, Labanotation comprises a group of symbols which, when combined, represent four parameters of dance steps:[2]

1. Direction
   - high
   - middle
   - low
2. Part of the body
3. Duration
4. Dynamic quality
   - Space (direct or indirect)
   - Weight (strong or light)
   - Time (sudden or sustained)
   - Flow (bound or free)

Each category has its own symbols, which, when arranged together, describe the movements to be performed by a dancer.[3]

During the 1930s–1950s, American-born dancer, choreographer, and theorist Ann Hutchinson Guest studied Laban's notation keenly in Europe and systematised his theories. In addition to milestone publications such as *Labanotation* (1954), *You Move* (1983), *Dance Notation* (1984), and *Choreographics* (1998), Guest funded the Dance Nation Bureau (DNB) in New York in 1940 and the Language of Dance Centre (LOD) in London in 1967, both centres focussing on teaching kinetography and recording choreographies throughout the world using Labanotation.

As mentioned in Part I, since the 1960s and 70s, critics have paid much attention to the conception of dance as language thanks to the so-called corporeal or body turn, which has benefited from research being done in numerous fields, including neurosciences and psychology. Therefore, while Laban and his disciples established a symbolic convention of signs to represent dance movements, even resorting to the principles of such hard sciences as geometry and physics, the corporeal turn demands that scholars focus on those internal, cerebral mechanisms that connect dance movements and verbal language. In the words of Judith L. Hanna, probably one of the finest researchers in the field,

> Research has shed light on how dance is similar in many ways to verbal language [...]. [D]ance draws upon the same components of the brain for conceptualization, creativity, and memory as verbal language. Both dance and verbal language have vocabulary (steps and gestures in dance), grammar (rules for justifying how one

> dance movement can follow another), and meaning. Although spoken language can simply be meaningless sounds, and movements can be mere motion, listeners and viewers tend to read meaning into what they hear and see. Both verbal language and dance also have arbitrariness (many of their characteristics have no predictability), discreteness (separateness), displacement (reference can be made to something not immediately present), productivity (messages never created before can be sent and understood within a set of structural principles), duality of patterning (a system of physical action and a system of meaning), cultural transmission, ambiguity, affectivity (expression of an internal state with the potential for changing moods and situations), and a wide range in the number of potential participants in their communication process. There are countless other intersections between dance and verbal language (2001, 41).

In other words, as Henrietta Bannerman has recently stated (2014), both language and dance are based on the same common denominator: communication. Speakers/writers and dancers/choreographers initiate their actions because they are moved by the need to communicate something to their listeners, readers, audience, etc. Despite acknowledging unbridgeable differences between dance and language,[4] Bannerman believes that "dance is structured in a similar way to language and that categories formulated within linguistic theory are commensurate with ways in which dances are constructed". Indeed, "*vocabulary* and *syntax* are present in dance in the way that the word vocabulary is often employed to describe the selection of specific movements, and syntax, to represent the combination or arrangement of these movements (vocabulary) into chains or phrases of dance material" (66, emphasis in the original).

When it comes to more complex structures such as speech acts or sentences, however, Bannerman is sceptical and affirms that they are not representable in dance. This must certainly be true in the case of early modern English dances inserted in sixteenth- and seventeenth-century plays, where in most cases choreographies accompanied and mirrored the characters' turn takings – thus often subordinating bodily movements to the lines uttered. Nevertheless, I believe that when dance is the only protagonist on stage, in the absence of verbal language, it is connoted by a strong tendency towards dialogism – in a purely Bakhtinian sense – and storytelling (see Ciambella 2017). Drawing on Walter Sorell's structuralist idea of "dance narrative" (1967),[5] the Bulgarian semiotician Smilen Antonov Savov, a member of the UNESCO International Dance Council, explored the narrative potential of dance and came up with the theorisation of an out-and-out narratology of dance, stating that "dance is the only art form that uses complex sign language to portray 'words', phrases, fragments, metaphors, language sequences, sentences that directly affect human emotions". Moreover, "[t]he dance performance is a semiotic construct of signs and semiotic formations, and is basically a defined set of codes combined with the secondary sign systems, such as mythology, literature, fine arts, music, and religion" (2014, online). Therefore, in addition to "grammar" rules, which allow dancers and choreographers to communicate with their audience effectively, the terpsichorean art has its own narratology, i.e., its own way of telling stories.

Many professional dance centres and projects have focussed recently on the intermingling of dance and storytelling, each centre/project acknowledging that dancing equals recounting stories. Among them the Peggy Baker Dance Projects (https://peggybakerdance.com/) is certainly one of the most famous and important charitable non-profit organisations whose mission revolves around storytelling. Established in Toronto, Canada, in 1990, the centre has four different performances in its storytelling repertoire: *A True Story*, *Home*, *The Disappearance of Right and Left*, and *Unmoored*.

Another noteworthy non-profit organisation is the Synetic Theater,[6] founded in Arlington, Virginia, in 2001 by the Georgian-born couple Paata and Irina Tsikurishvili. The Synetic Theater has produced numerous wordless adaptations of Shakespeare's plays (the latest, *King Lear*, marking the company's 20th anniversary in 2021), transforming the Bard's lines into dumb shows and thus expressing dance narratology at its best (see Ciambella 2017).

Similarly, at the end of 2019, born of a collaboration between the International Writing Program and the Department of Dance at the University of Iowa, the *Words Dance* showcase was launched. Five authors of the International Writing Program had the privilege of assisting the transmedialisation of their works of fiction into choreographies and the success of the initiative was astonishing. Compared to other adaptations of written works into dance performances – e.g., Prokof'ev's *Romeo and Juliet* (1938) – this initiative might not seem that innovative; however, the novelty of the *Words Dance* showcase is that the author of the written source text collaborates closely with choreographers and dancers, reading his/her work aloud and, more importantly, deliberately writing a narrative that must be adapted for a terpsichorean performance.

Thus the above-mentioned studies and projects have demonstrated that similarities between language and dance can be found at almost any level of analysis – i.e., dance vocabulary, syntax, narrative, etc. –

and that a wide range of sciences and research fields have contributed to identifying such connections between natural languages and the terpsichorean art. Before concluding, one should mention some of the latest pivotal studies which have focused on the application of linguistic – especially cognitivist – methodologies to dance, in order to anticipate some of the methodological issues presented in the second half of this part.

Acknowledging that "both [language and dance] have a biological foundation [...] and the body is their expressive tool" (Napoli and Kraus 2017, 468), the linguist Donna Jo Napoli and the professional dancer and educator Lisa Kraus aimed at tracing a parametric typology of dance which does not consider cultural issues at all, something that, for obvious reasons, is not possible in this book. Their assumption is that exactly as language typology allows linguists to reveal similarities among typologically distant languages, which transcend cultural proximity between two or more of these languages, even dance can be analysed using the culture-free parameters of typology. Hence, a parameter-based typology of dance is offered here and typologically-distant styles of dances are compared and contrasted according to ten parameters:

1. Prominence (side, direction and facing)
2. Alignment (vertical/horizontal and orientation)
3. Gravity
4. Inversion
5. Space
6. Sequencing
7. Rhythm
8. Initiator
9. Quality of Movement
10. Tenseness

For example, considering the parameter of alignment, Napoli and Kraus managed to find typological similarities between Cambodian, West African, Release, and Hip Hop since feet are parallel, unlike Modern and Contemporary dance, where foot position varies, and Ballet with its turned-out feet positions.

A cognitive semantic approach characterises Patel-Grosz et al.'s analysis of co-reference and disjoint reference, typical of noun phrases, in dance, with the long-term goal of "establish[ing] the common semantic properties of (non-)linguistic cognitive systems" (2018, 199). In their case, the non-linguistic cognitive system is that of a particular dance narrative of Indian origin, the Bharatanatyam. In line with the notion of grouping coined by Lerdhal and Jackendoff in relation to tonal music (1983) and Abusch's semantic of visual narrative (2013; 2014; 2020[7]), Patel-Grosz et al. hypothesise that "dance shares hierarchical grouping with language and music, which may be interpreted as giving rise to a syntactic structure of dance" (2018, 206).

Lastly, resorting to the same Chomskian/generativist-derived concept of grouping also used by Patel-Grosz et al., Isabelle Charnavel, of the Department of Linguistics at Harvard University, conducted an experiment based on the replication of some arm movements in order to demonstrate "that dance obeys grouping principles, just like other cognitive domains such as language (esp. prosody), music or vision" (2019, 15),[8] so that a universal grammar of dance may be possible. The parameters used by Charnavel to conduct her experiment do not differ that much from those identified by Napoli and Kraus listed above (2017):

1. Configuration (body part)
2. Weight (which part of the body supports the dancer's weight?)
3. Orientation of the dancer
4. Direction
5. Speed
6. Quality of the movement

Without detailing the results of these latest studies, what is important to conclude about them is that in recent years typology-based or generativist research on the intersections between language and dance tend to focus on the detection of universal parameters which can associate different dance styles. These contemporary approaches can be adopted when analysing early modern dance in sixteenth- and seventeenth-century English plays when seeking a universal semantics of Renaissance dance lexis, one that also focuses, however on chronotopic features typical of that culture, since, I would argue, the issues of corpus and textual linguistics outlined in the next sections cannot overlook culture-bound aspects.

Notes

1 One of the ancestors of Labanotation is the baroque Beauchamp-Fueillet notation, invented by Pierre Beauchamp in 1680 under the commission of Louis XIV of France and described for the first time in detail by Raoul-Auger Feuillet in his *Choréographie* (1700). In the nineteenth century, both French and Russian dancing masters and choreographers elaborated their dance notations: Arthur Saint-Léon invented his Sténochorégraphie in 1852,

while Vladimir Stepanov outlined his theories about dance stenography in his *L'Alphabet des mouvements du corps humain* (1879). More importantly, the Benesh Movement Notation (BMN), invented by Joan and Rudolph Benesh in the late 1940s, anticipated Laban's kinetography. This is the dance notation still taught and learnt at the Royal Academy of Dance in London (see https://www.royalacademyofdance.org/benesh-international-benesh-movement-notation/).

2 The basics of Labanotation can be studied further at the Dance Nation Bureau website (http://dancenotation.org/) or downloaded as a PDF at the Language of Dance Centre website (https://www.lodc.org/old-about-us/what-is-language-of-dance/the-movement-alphabet.html).

3 For further information about Labanotation, see Pforsich 1977; Hutchinson Guest 2005.

4 See also Margolis 1983 about differences between language and dance.

5 It is worth noting here Sorell's pivotal contribution to the study of Shakespearean dances (1957), even prior to Brissenden's *Shakespeare and the Dance*.

6 The word "synetic" is the partial blend of synthesis and kinetic, meaning a dynamic synthesis of different arts – almost exclusively verbal language and dance.

7 The authors refer to the 2015 draft version of this article published in 2020, two years after Patel-Grosz et al.'s 2018 contribution.

8 Actually, Charnavel's article was quoted repeatedly by Patel-Grosz et al. in its embryonic 2016 manuscript version.

6 Language *about* dance: Matters of corpus and textual linguistics

The second approach I would like to examine, which is typical of corpus and textual linguistics and of stylistics, is the one that this book has benefited from the most. As Susan L. Wiesner (2011, 2) has noted, none of the dance notations examined thus far completely covers all the meanings and forms of dance; hence, experts and scholars must fill this communicative gap with forms of dance writing. I believe that gap-filling attempts to find a language *about* dance that has contributed to the rise and dissemination of a terpsichorean LSP (Language for Special/Specific Purposes) in most of the natural languages, including English. Offering a valid contribution to the analysis of dance ESP (English for Special/Specific Purposes), Wiesner, while a PhD student at the University of Surrey, created a corpus – which she named the Surrey Dance Corpus – comprising miscellaneous materials about dance (journal/newspaper articles, manuals, Internet archives, etc.) to be investigated using corpus linguistic tools (KonText[1] in this case). By creating subcorpora for different textual genres (reviews, articles, reports, etc.), Wiesner studied the most recurring words in the corpus, their concordances and different patternings using the KWIC (KeyWords In Context) function, which is precisely what will be done in the next part of this book by selecting a completely different corpus and more advanced corpus linguistic tools.[2]

A similar approach has been adopted by Mirosława Podhajecka, from the Department of English at Opole University, Poland (2010; 2011). Podhajecka tried to demonstrate that the Google Books collection can improve the results obtained with more traditional corpora, such as those used to edit and publish the *Oxford English Dictionary*.[3] Although not being a dance expert and using dance-related terms simply as a case study, "since dances have always been a vital element of cultural and social histories of nations" (2010, 1045), the data-driven methods she employed are particularly useful because she very lucidly outlines the series of limits and issues connected to the selection of material and the results obtained.

Speaking of the *OED* initiative, a corpus-driven approach to a dance lexicon led to the publication of the *Oxford Dictionary of Dance* in 2000 (second edition, 2010), perhaps the most complete dictionary of terpsichorean terminology ever compiled, with over 2,600 entries and definitions. In addition to dance styles, famous choreographers, and important twentieth- and twenty-first-century dance events, the *Oxford Dictionary of Dance* provides lemmas and definitions of difficult technical terms and dance names, not to mention hints about dance notations.

The studies mentioned above are probably the only data-driven contributions focussing on language *about* dance or dance writing. Nevertheless, long before approaching dance jargon via corpus linguistics, some scholars understood the importance of surveys about terpsichorean terminology in dance-related texts, especially when pedagogical implications about dance teaching are present. In other words, these textual linguistic studies focussed on the necessity of what I would call a "core vocabulary of dance" allowing for effective communication between dancing masters/choreographers and their pupils.

Given that "dancing is the oldest of all the arts" (1916, 9), in that dance vocabulary passed from language to language almost unconsciously, Professor Frederick A. Wright lamented the lack of dictionaries about dance technicalities. Far from being an outright article on lexicography, Wright's contribution was a philologically accurate survey aimed at tracing the origin – mainly from ancient Greek – of some generic terms which underwent a semantic turn and meaning restriction, and entered the lexis of dance as specialised vocabulary.

Although French and Russian dictionaries of dance-related technical vocabulary began to appear in the eighteenth and nineteenth centuries (e.g., Charles Compan's *Dictionnaire de la dance*, 1787, and its Russian translation, *Танцевальный словарь*, in 1790; Gustave Desrat's *Dictionnaire de la danse*, 1895), the first complete dance dictionary in English was Dorothy N. Cropper's *Dance Dictionary*, published in New York in 1935, most likely because, as explained below, it was only at the beginning of the twentieth century that dance ESP began to develop independently from French (and Italian) models.

In 1992, examining eighteenth- and nineteenth-century dance manuals, Sandra N. Hammond demonstrated that dance lexis in English was still vastly influenced by French and Italian terpsichorean vocabulary – as the analysis of dance lexicon in early modern English plays presented in the next section of this book confirms. Although Playford's *The English Dancing Master* represented a turning point towards the "nationalisation" of the English Country Dance and terpsichorean vocabulary in English as well, French and Italian terpsichorean lexis continued to influence dance ESP for a few centuries, at least, I would argue, until modern dance was born in the United States (thus in an anglophone environment) with dancers and choreographers such as Isadora Duncan and Martha Graham, among others. It is no accident that the adjective "English" disappeared from the title of Playford's manual immediately after the first edition in 1651.

Last, but not least, Akiko Yuzurihara (2005; 2013) reflected on the fact that the creation and experimentation of a language *for* dance contributed to the rise of a language *about* dance. In other words, to make such dance notations as kinetography intelligible, dance theorists had to express and explain them with words, thus creating a kind of specialised metalanguage of dance, a language *about* dance used to define the language *for* dance. In this way, informative-expository texts such as Guest's *Labanotation* become precious sources for a metalinguistic analysis of dance ESP.

Notes

1 KonText (Knowledge ON TEXT) is one of the oldest corpus linguistic analysis tools, developed at the University of Surrey, which allows one to process texts and perform simple operations such as wordlists and KWIC (KeyWord In Context). Although its graphic interface is quite old-fashioned and obsolete, the results provided are interesting and significant.
2 Wiesner's ideas and research findings were updated in a 2019 article about dance ESP.
3 Authors of the *OED* use the so-called Oxford English Corpus, a collection of more than 2 billion words.

7 Early modern English lexicography: Limiting the scope

As studies about early modern English dance lexicons seem quite rare, this book has benefited also from more general research about sixteenth- and seventeenth-century lexicography which has contributed to the methodological organisation of the analysis presented here. When dealing with early modern English lexicography, the majority of studies take into account the 1600s attempts to publish monolingual dictionaries of the English language, "from Robert Cawdrey's *A Table Alphabeticall* (1604) onwards" (Iamartino 2017, 56). Nevertheless, to speak about dance lexis and its continental influences, sixteenth- and seventeenth-century bilingual dictionaries such as Claude Desainliens's *A Dictionary French and English* (1593) or John Florio's Italian-English dictionaries (*A World of Words*, 1598; *Queen Anna's New World of Words*, 1611) are worth examining. Given their pedagogical stamp (cf. Elam 2007, 120; Montini 2015), these bilingual dictionaries were produced "in response to very practical needs" (Osselton 1983, 13) and "must [...] be judged from the perspective of their users" (Morini 2020, 18), who in early modern England were mainly gentlemen and gentlewomen. Those very same gentlemen and gentlewomen who used to learn foreign languages such as French or Italian by resorting to the bilingual dictionaries of Desainliens and Florio were the dedicatees and readers of early modern terpsichorean manuals and treatises – for conduct books recommended dancing as physical exercise. Thus it is customary for sixteenth- and seventeenth-century bilingual French/Italian-English dictionaries to contain many lemmas concerning dance names, steps, movements, etc. (for an analysis of dance-related definitions in Florio's 1611 dictionary, see, inter alia, Sándorfia 1996).

First published in 1989, the two volumes of *Early Modern English Lexicography* are clearly one of the primary sources for lexicographers conducting research on this period of English history. While volume 1 is conceived as a census of glossaries and dictionaries from 1475 to 1640, volume 2 provides the definitions of a series of lemmas not included in the *OED*. Schäfer's editorial enterprise represents a specialised preliminary analysis of early modern lexis which also considers spelling variations.

On the basis of *Early Modern English Lexicography* and drawing on recent studies on the use of corpora in lexicography (see, i.a., Zampolli 1983), digital databases of early modern English lexis began to appear.[1] Ian Lancashire, from the Department of English at University of Toronto, began researching corpus-based dictionaries of early modern English in 1992, and in 1996 he created the Early Modern English Dictionaries Database (EMEDD), which lemmatised almost 200,000 thousand words from 1530 to 1657 whose definitions were taken from 16 different works:

- six bilingual dictionaries: John Palsgrave (1530; English-French), William Thomas (1550; Italian-English), Thomas Thomas (1587; Latin-English), John Florio (1598; Italian-English), John Minsheu (1599; Spanish-English), and Randle Cotgrave (1611; French-English). These cover four other languages and give pairs of French and Italian separated by 50–80 years.
- five English hard-word dictionaries: Edmund Coote (1596), Robert Cawdrey (1604, based on the transcription by Raymond Siemens; and 1617), John Bullokar (1616), and Henry Cockeram (1623).
- the first full English-only dictionary, by Thomas Blount (1656).
- three specialized lexicons: Bartholomew Traheron's translation of Vigon (1543), William Turner on herbal names (1548), and John Garfield on scientific terms in J. Renou's *Dispensatory* (1657).
- the first full English word-list by Richard Mulcaster in his *The first part of the Elementarie* (1582).

(Lancashire 1999, online)

In 2000, the EMEDD was replaced by the Lexicons of Early Modern English project (LEME, available at https://leme.library.utoronto.ca/), which provides the definitions of 500,000 entries from 1480 to 1755. LEME is a much more complete and complex project, "a historical database of monolingual, bilingual, and polyglot dictionaries, lexical encyclopedias, hard-word glossaries, spelling lists, and lexically-valuable treatises surviving in print or manuscript" (Lancashire 2018, online), whose graphic interface is also easier and more intuitive than EMEDD's. It enables quick and advanced searches, in addition to the modern English headwords search with normalised lemmas, the proximity search to combine two words and their left/right concordances, and the Boolean lexicon search. Moreover, LEME allows one to order various lexicons by date, title, author, subject, and genre, and to download them.

Note

1 One of the main pioneers of corpus linguistics, Charles C. Fries, had already envisaged the possibility of a corpus-based dictionary of early modern English when becoming editor of the Early Modern English Dictionary (EMED) in 1928, a position he held until 1958 (Fries 2010).

8 Corpus selection and investigation

The final section of Part II deals with the corpus of early modern English plays selected and the corpus linguistic tool chosen for the lexical analysis presented in the next chapter.

The VEP Early Modern Drama Collection

As previously noted, corpus linguistics has been chosen as the privileged methodology to investigate dance lexis in early modern English plays, given its many advantages to this kind of analysis.[1] As Jane Demmen has recently pointed out about the Encyclopedia of Shakespeare's Language Project (ESLP)[2] at the University of Lancaster (2020), the main difficulty in compiling a corpus of early modern English plays, aside from chronological matters related to the label "early modern" itself,[3] stems from William Shakespeare's hegemonic position within the sixteenth- and seventeenth-century literary milieu:

> However, Shakespeare was just one among a number of well-known and successful playwrights writing in the late 16th and early 17th centuries. In the field of corpus linguistics that explores language style features (corpus stylistics), relatively little attention has thus far been given to investigating the language of playwrights other than Shakespeare. [...] Quantitatively-based comparisons of Shakespeare's language style relative to those of his peers are to date mainly restricted to computational stylistic research, focusing on authorship attribution [...]. Areas as yet largely unexplored include pragmatic phenomena, style features of dramatic sub-genres, metaphor use, and characterisation of people of different gender and social rank through language style (38).

In the analysis presented in this book, Shakespeare's canon is examined in tandem with some of the most dance-aficionado contemporaries of the Bard, because works such as Marston's plays or Jonson's masques can certainly offer outstanding views of the lexicography of dance in early modern England. For this reason, the selection of the corpus to be investigated requires particular attention and the choices made must be justified.

Of course, when dealing with early modern English textuality, the most valuable source is EEBO-TCP (available at https://quod.lib.umich.edu/e/eebogroup/). With its more than 60,000 digitalised texts between phase I (25,368) and phase II (34,963) thus far, the Early English Books Online repository is the largest archive of miscellaneous texts[4] in English – with some Latin and French writings available as well – written and published from 1475 to 1700. Due to the disruptions provoked by the COVID-19 pandemic, both phase I and phase II texts were made freely available to any user as of 1 August 2020. Given the incredibly wide range of textual genres available, EEBO cannot be a representative and statistically-significant corpus for the study carried out in this book about drama. Nevertheless, as best practice in corpus linguistics, I resort to EEBO as a reference corpus, i.e., a large, general corpus providing comprehensive information about a language (in this case in a specific period of the British history).

Conversely, focussing exclusively on sixteenth- and seventeenth-century English plays, A Digital Anthology of Early Modern English Drama (EMED) could be a significant corpus to be considered here. EMED (available at emed.folger.edu) was created between 2015 and 2017 by the Folger Shakespeare Library with the specific aim of digitising plays written by Shakespeare's contemporaries[5] from 1576 (when the first Elizabethan playhouse, The Theatre, was built in London) to 1642 (when the Civil War broke out and public theatres were closed in England). This collection of 403 plays can be consulted directly at the EMED website. The interface has an internal search tool through which users can search for keywords, author and/or title of the play, and even restrict the search if required to genre, company and/or theatre. Although some 29 plays – both in original and regularised spelling – are actually downloadable in HTML, PDF, or XML format, the analysis of the remainder of the corpus is limited to the search tool available at the website which, aside from the keyword extraction, cannot perform the advanced operations that corpus linguistic software such as #Lancsbox, Sketch Engine, or AntConc (to mention only a few) can carry out, as introduced in the next section. For this reason, EMED has not been taken into account in this book.

It was the Visualizing English Print (VEP) Early Modern Drama Collection that contributed the most to the corpus selection and exploration I conducted.[6] VEP Early Modern Drama Collection is part of the VEP project, a Mellon-funded endeavour begun in 2016 in collaboration with the University of Madison-Wisconsin, the University of Strathclyde (Glasgow, Scotland), and the Folger Shakespeare Library. The mission of the project is "to scale humanist scholarship to 'big data' by removing several barriers to statistically analysing digital Early Modern English texts, like those released by the Text Creation Partnership". What is more, they "offer plain text corpora of Early Modern English texts and visualization tools to explore them. We also provide the text processing pipeline used to curate our corpora. This website equips you with the necessary tools to make your own corpora, visually explore them, and download preliminary metadata" (Stuffer 2016, online). Members of the VEP project are currently working on the creation of five different corpora, among them the Early Modern Drama Collection, which is the one that benefits from the contribution of Professor Jonathan Hope. It is organised into three main subcorpora: Core Drama 1660, Expanded Drama 1660, and Expanded Drama 1700. The Core Drama 1660 corpus is a collection of 554 plays, including professional and other plays intended for performance, translations from plays from other languages, and closet drama until 1660. The Expanded Drama 1660 corpus collects 666 plays written and performed until 1660, with the addition of masques and entertainments, while the Expanded Drama 1700, which includes 1,244 plays, also contains works belonging to the Restoration drama until 1700. For the purpose of this study, the Expanded Drama 1660 subcorpus has been chosen and plain text files downloaded, together with the wordlist of the entire corpus with word forms ordered by frequency. As far as normalisation of early modern English spelling is concerned, the VEP team opted not to use the VARD 2[7] software developed at the University of Lancaster, as it would have been too time-consuming. Instead, they decided to develop their own dictionary of spelling standardisation which, however, left some morphological aspects of sixteenth- and seventeenth-century English unaltered – e.g., the second person singular inflection -(e)th of the present tense.

Lastly, since only EEBO-TCP phase I texts are available for download, phase II texts have been identified by comparing the list of downloadable texts to that of the Expanded Drama 1660 and downloaded directly by EEBO. As 569 plays out of 666 belong to EEBO phase I, 97 plays have been downloaded and then normalised using VARD 2 software as plain texts before uploading the entire corpus on

#Lancsbox.[8] For the sake of convenience, in this study I indicate this corpus of 666 digitalized texts as the Early Modern English Plays (hereafter EMEP) corpus.

The #Lancsbox software and lexical analysis

Among the numerous corpus linguistic software available, I have decided to use #Lancsbox. I explain the reasons for this choice and illustrate the main functions related to lexical analysis below.

#Lancsbox (the Lancaster University corpus toolbox), a very intuitive and efficient software with which to explore corpora, is downloadable at http://corpora.lancs.ac.uk/lancsbox/. It was developed at the University of Lancaster in 2015 (see Brezina et al. 2015) by Vaclav Brezina and Professor Tony McEnery with the aim of helping linguists, lexicographers, teachers, and students understand how language works, what is typical, and what is rare and unique. At the moment, #Lancsbox supports both left-to-right and right-to-left languages, but it can annotate parts of speech for 20 of them.[9] It also contains some large and representative corpora – even multilingual and parallel ones – in different languages: BNC (British National Corpus), Brown, LCMC (Lancaster Corpus of Mandarin Chinese), LOB (Lancaster-Oslo-Bergen), Newsbooks, Shakespeare, and VULC (Vrjie Universiteit Learner Corpus). Some of these corpora contain up to 100 million words. The main advantages #Lancsbox offers in the field of specialised lexicography are listed in the introductory page of its website:

#LancsBox is a new-generation software package for the analysis of language data and corpora developed at Lancaster University.

Main features of #LancsBox:

- Works with your own data or existing corpora.
- Can be used by linguists, language teachers, historians, sociologists, educators and anyone interested in language.
- Visualizes language data.
- Analyses data in any language. [...]
- Automatically annotates data for part-of-speech.
- Works with any major operating system (Windows, Mac, Linux).

(http://corpora.lancs.ac.uk/lancsbox/index.php)

Thus to the complex functions and algorithms of the software correspond a user-friendly graphic interface and a just-one-click functioning, which make the #Lancsbox software highly intuitive and easily accessible.

#Lancsbox gives its users the possibility of selecting one of the corpora in its repository or of creating their own corpus by choosing among the many languages it reads. When creating a corpus, before uploading the single texts in.txt,.pdf, etc., the software asks for a title and the language(s) of the corpus. I have created a corpus named Early Modern English Plays (EMEP) and #Lancsbox automatically tagged the various parts of speech before letting me work directly with its seven main functions,[10] the most useful of which I will now briefly describe.

The most important and useful function available from the software dashboard for lexicographic analysis is no doubt the KWIC tool, which provides co-textual information about the token[11] under scrutiny. Once a word is searched, its occurrences are shown in the statistical analysis bar together with the number of texts in the corpus which contains that word. The bar also allows the #Lancsbox user to set the number of right and left collocates to be shown in the search. The results of the KWIC search are listed in order of appearance in the concordance display, in a single raw with the node – i.e., the token looked for – in the middle and the selected number of left and right context on both sides. Results can be shown as plain text, but also as parts of speech or lemmas. Double clicking on the node opens a pop-up window with the largest portions of the text where the word has been detected in order to provide an even broader context. Concordances can also be saved and downloaded as.txt files.

Another important tool from which this study has benefited is Words. As mentioned earlier, the VEP Early Modern Drama Collection already has lists of words ordered from the most frequent to hapaxes, but such terms are not tagged. In other words, there is no distinction, for instance, between the noun "dance" and the verb "to dance", because parts of speech are not signalled. The "Words" function in #Lancsbox, however, allows users to seek words belonging to one specific grammatical class, or to distinguish between type, lemma, and part of speech by simply clicking on the "type" function.

In addition, the Whelk tool has proved useful for the analysis conducted in Part III. It is a function which provides information about the distribution of the node sought in the various corpus texts. This tool has been particularly useful in locating those plays listing the highest number of dances (either performed or simply mentioned), thus helping to put forward hypotheses about the possible qualitative interpretations of the data extracted (e.g., Is the relative frequency of dance-related lemmas highest in such particular genres as masques?).

Lastly, GraphColl has been taken into account. This tool provides information about the collocational patterning of the node searched, thus visualizing both left and right collocations in a collocation network graph and their frequency of combination with the word observed. Results are visualized according to three parameters:

1. Strength: The closer a collocation is to the node, the stronger its association with the node.
2. Frequency: The darker the colour of a collocation is, the more frequent its association with the node.
3. Position: It can be left, right, or in the middle, if the collocate is found both on the left and on the right of the node.

Moreover, the GraphColl function can find the collocates that two different lemmas have in common. This function, for example, is very important when comparing two dances such as the Roundlet or the Ringlet, which can be considered synonyms, but whose collocational patterning might differ for stylistic reasons.

Which words to look for

Once the EMEP corpus has been created, #Lancsbox needs inputs to perform its functions. These inputs are the words, word forms, lexemes, and lemmas to be searched. Hence, which words should we look for, considering that an out-and-out early modern dance dictionary does not exist?[12] To deal with this issue, two different but complementary sources have been adopted, in order to obtain a list of early modern dances – both courtly and popular – covering most of the terpsichorean lexis of the period taken into consideration.[13]

On the one hand, Jim Hoskin's *The Dances of Shakespeare* (2005) provides a basic list of dances the Bard mentioned and/or included in his canon. These dances are (in alphabetical order):[14]

- Bergomask
- Branle/Brawl
- Canary
- Coranto/Courante
- Galliard
- Hay
- Jig/Gig/Gigue

- (La)Volta
- Maypole Dance
- Measure
- Morris Dance
- Pavan
- Round(el)/Ring(let)

Moreover, the choreographer dedicates two chapters to stage directions where the lexemes "dance" and "masque" appear.

On the other hand, drawing on the 16 monolingual and bilingual early modern dictionaries available at the EMEDD/LEME project, Greg Lindhal (2018) extracted the dance-related terminology and highlighted both the similarities and differences among those dictionaries. Although, unlike Hoskin's book, Lindhal's article is not based on sixteenth- and seventeenth-century plays, but on dictionaries, the results obtained, and the dances and steps listed, may broaden our perspective on early modern dance lexicography by also considering choreographies that Shakespeare did not include in his plays. Some of the additional dances listed in most dictionaries and which can be expected to be found in the EMEP corpus are the following:

- Almain/Allemand(e)
- Cinquepace/Sinkapace
- Country Dance
- Cushion Dance
- Horn(i/e)pipe
- Moresca/Morisco
- Passamezzo

The analysis presented in the next chapter begins by focussing on the above-mentioned dances. Entries are listed in alphabetical order – as a kind of "annotated glossary" of early modern dances – and each choreography is briefly explained by providing the definition offered mainly by the *Oxford Dictionary of Dance* (2010, second edition) and the *Oxford Dictionary of Music* (2013, sixth edition), and only in two cases – i.e., Cinquepace/Sinkapace and Maypole Dance – respectively by the *Oxford Companion to Shakespeare* (2015, second edition) or the *Encyclopædia Britannica*. Examples from the EMEP corpus are then provided and commented on throughout the qualitative analysis presented.

Notes

1 The literature on the astonishing contribution that corpus linguistic tools can offer lexicographic and lexicological analyses is vast. For some chronologically-distributed references, see, among others, Teubert 2001; Williams 2003; Lauder 2010; Gizatova 2016 and Brezina 2018, esp. 38–65, 102–138.
2 Visit http://wp.lancs.ac.uk/shakespearelang/ for further information.
3 Demmen, following Nevalainen (2006, 1), adopts the time span 1500–1700 in her article (2020, 62).
4 The EEBO website (https://textcreationpartnership.org/tcp-texts/eebo-tcp-early-english-books-online/) reads: "The books in these collections include works of literature, philosophy, politics, religion, geography, history, politics, mathematics, music, the practical arts, natural science, and all other areas of human endeavour".
5 Shakespeare's works are available at the Folger Shakespeare Library website at https://shakespeare.folger.edu/; for that reason the EMED project did not consider the Bard's output.
6 At this juncture, I would like to "officially" thank the members of the executive board 2018–21 of the Italian Association of Shakespeare and Early Modern Studies (IASEMS) – namely, in alphabetical order of surname, Dr. Luca Baratta, Prof. Maria Luisa De Rinaldis, Prof. Gilberta Golinelli, Prof. (and President) Giuliana Iannaccaro, and Prof Iolanda Plescia – and my dear friend and colleague Domenico Lovascio for organising the 10th IASEMS conference in Genoa (22–24 May 2019), where I had the opportunity to meet Prof. Jonathan Hope and first hear about the VEP Early Modern Drama Collection during his plenary.
7 VARD 2 "is an interactive piece of software produced in Java designed to assist users of historical corpora in dealing with spelling variation" (http://ucrel.lancs.ac.uk/vard/about/). It was developed by the Lancaster University Centre for Computer Corpus Research on Language (UCREL).
8 VARD 2 "uses techniques derived from modern spell checkers to find candidate modern form replacements for spelling variants found within historical texts. The user can choose to process texts manually [...] or semi-automatically" (http://ucrel.lancs.ac.uk/vard/about/). In this case, following the same set of "rules" used by the VEP dictionary of spelling standardisation, I processed the 97 texts manually.
9 Arabic, Catalan, Czech, Dutch, English, Finnish, German, Italian, Latin, Mongolian, Portuguese, Romanian, Russian, Slovak, Spanish, and Swahili.
10 KWIC, GraphColl, Whelk, Words, Ngrams, Text and Wizard.
11 "An individual occurrence of a linguistic unit in speech or writing" (*OED* n. 3).
12 The closest publication to the aims of this analysis is probably Joseph P. Swain's *Historical Dictionary of Baroque Music* (2013), whose focus, however, is different from the one presented here.
13 The names of some steps and titles of choreographies are provided in the analysis presented in Part III, but they are not reported as separate entries of the glossary for various reasons. As critics point out, the complex steps described by Caroso, Arbeau, and Negri were too sophisticated and

elaborate to be of interest to English playwrights, who were certainly not professional dancers. In fact, sixteenth- and seventeenth-century English manuscripts "contain no step descriptions" (Early Dance Circle, online). As I argue elsewhere (Ciambella 2020, 39–41), even Playford's manual focuses more on directions (left, right, forward, and backward) and repetitions (single and double) than on the specific names of the steps to be performed. Moreover, as Winerock has noted (2016, online), there are a few cases when mentioning the technical names of some steps indicated an elitist and aristocratic behaviour of superiority by one character towards the other(s) on stage, as happens in the dialogue between Sir Toby and Sir Andrew in Shakespeare's *Twelfth Night* 1.3.

14 The spelling of these dances varied considerably during the sixteenth and seventeenth centuries, due both to the highly fluctuating spelling of early modern English and the difficulties arising from the French/Italian origin of some choreographies. Therefore, in this study, spelling variations are indicated whenever necessary and not immediately apparent.

References

Abusch, D. (2013) Applying discourse semantics and pragmatics to co-reference in picture sequences. *Proceedings of Sinn und Bedeutung* 17, 9–25.

Abusch, D. (2014) Temporal succession and aspectual type in visual narrative. In: Crnič, L. and Sauerland, U. (eds.) *The Art and Craft of Semantics: A Festschrift for Irene Heim*. Vol. 1. Cambridge, MA: MIT Press, 9–20.

Abusch, D. (2020) Possible worlds semantics for pictures. In: Matthewson, L., Meier, C., Rullmann, H., and Zimmermann, T.E. (eds.) *The Wiley Blackwell Companion to Semantics*. Oxford: Wiley.

Bannerman, H. (2014) Is dance a language? Movement, meaning and communication. *Dance Research: The Journal of the Society for Dance Research* 32(1), 65–80.

Brezina, V., McEnery, T. and Wattam, S. (2015) Collocations in context: A new perspective on collocation networks. *International Journal of Corpus Linguistics* 20(2), 139–173.

Brezina, V. (2018) *Statistics in Corpus Linguistics: A Practical Guide*. Cambridge: Cambridge University Press.

Charnavel, I. (2019) Steps toward a universal grammar of dance: Local grouping structure in basic human movement perception. *Frontiers in Psychology* 10, 1–19.

Ciambella, F. (2017) Dalla parola al *dumb dance show*: *Twelfth Night* del Synetic Theater. In: Tempera, M. and Elam, K. (eds.) *Twelfth Night. Dal testo alla scena*. Bologna: ELIM, 205–220.

Ciambella, F. (2020) The [Italian] dancing master: English reception of Italian Renaissance terpsichorean manuals. A corpus-driven analysis. In: Magazzù, G., Rossi, V. and Sileo, A. (eds.) *Reception Studies and Adaptation: A Focus of Italy*. Newcastle upon Tyne: Cambridge Scholars Publishing, 28–44.

Craine, D. and Mackrell, J. (2000) *The Oxford Dictionary of Dance*. Oxford: Oxford University Press.

Craine, D. and Mackrell, J. (2010) *The Oxford Dictionary of Dance*. Second edition. Oxford: Oxford University Press.

Demmen, J. (2020) Issues and challenges in compiling a corpus of early modern English plays for comparison with those of William Shakespeare. *International Computer Archive of Modern and Medieval English Journal* 44(1), 37–68.

Early Dance Circle. The Late Renaissance c1535–c1620. [Online]. Available at: https://www.earlydancecircle.co.uk/resources/dance-through-history/the-late-renaissance-c-1535-c-1620/. [Accessed on 31 October 2020].

Elam, K. (2007) "At the cubiculo": Shakespeare's problems with Italian language and culture. In: Galigani, G. (ed.) *Italomania(s): Italy and the English Speaking World from Chaucer to Seamus Heaney*. Firenze: Mauro Pagliai, 111–122.

Fries, P.H. (2010) Charles C. Fries, linguistics and corpus linguistics. *International Computer Archive of Modern and Medieval English Journal* 34, 89–120.

Gizatova, G. (2016) A corpus-based approach to lexicography: Towards a Thesaurus of English idioms. In: Margalitadze, T. and Meladze, G. (eds.) *Proceedings of the 17th EURALEX International Congress*. Tibilisi: Ivane Javakhishvili Tbilisi University Press, 348–354.

Hammond, S.N. (1992) Steps through time: Selected dance vocabulary of the eighteenth and nineteenth centuries. *Dance Research* 10(2), 93–108.

Hanna, J.L. (2001) The language of dance. *Journal of Physical Education Recreation & Dance* 72(4), 40–45.

Hoskins, J. (2005) *The Dances of Shakespeare*. London/New York: Routledge.

Hutchinson Guest, A. (2005) *Labanotation: The System of Analyzing and Recording Movement*. Fourth edition. New York/London: Routledge.

Iamartino, G. (2017) Lexicography, or the gentle art of making mistakes. *Other Modernities* (*Special Issue: Errors: Communication and Its Discontent*) 4, 48–78.

Kennedy, M. and Kennedy, J.B. (2013) *The Concise Oxford Dictionary of Music*. Sixth Edition. Oxford: Oxford University Press.

Lancashire, I. (1999) The Early Modern English Dictionaries Database (EMEDD). [Online]. Available at http://homes.chass.utoronto.ca/~ian/emedd.html. [Accessed on 19 August 2020].

Lancashire, I. (2018) Introduction to LEME. [Online]. Available at https://leme.library.utoronto.ca/help/intro. [Accessed on 19 August 2020].

Lauder, A.F. (2010) Data for lexicography: The central role of the corpus. *Wacana* 12(2), 219–242.

Lerdhal, F. and Jackendoff, R. (1983) *A Generative Theory of Tonal Music*. Cambridge, MA: MIT Press.

Lindahl, G. (2018) References to dance in sixteen early modern dictionaries. [Online]. Available at http://www.pbm.com/~lindahl/articles/dance_em_dict.html. [Accessed on 24 August 2020].

Montini, D. (2015) John Florio and Shakespeare: Life and language. *Memoria di Shakespeare: A Journal of Shakespearean Studies* 2, 109–129.

Morgolis, J. (1983) Art as language. In: Copeland, R. and Cohen, M. (eds.) *What is Dance?* Oxford: Oxford University Press, 376–387.

Morini, M. (2020) "Proper vnto the tongue wherein we speake": Robert Cawdrey's *Table Alphabeticall* and the archaizers. *Textus* 33(1), 17–34.

Napoli, D.J. and Kraus, L. (2017) Suggestions for a parametric typology of dance. *Leonardo* 50(5), 468–476.

Nevalainen, T. (2006) *An introduction to Early Modern English*. Edinburgh: Edinburgh University Press.

Osselton, N.E. (1983) On the history of dictionaries. In: Hartmann, R.R.K. (ed.) *Lexicography: Principles and Practice*. London: Academic Press, 13–21.

Patel-Grosz, P., Grosz, P.G., Kelkar, T. and Jensenius, A.R. (2018) Coreference and disjoint reference in the semantics of narrative dance. In: Sauerland, U. and Solt, S. (eds.) *Proceedings of Sinn und Bedeutung* 22(2), 199–216.

Pforsich, J. (1977) *Handbook for Laban Movement Analysis*. New York: Janis Pforsich.

Podhajecka, M. (2010) Antedating headwords in the third edition of the OED: Findings and problems. In: Dykstra, A. and Schoonheim, T. (eds.) *Proceedings of the XIV EURALEX International Congress (Leeuwarden, 6–10 July 2010)*. Leeuwarden: Fryske Akademy, 1044–1064.

Podhajecka, M. (2011) Research in historical lexicography: Can Google Books collection complement traditional corpora? In: Lewandowska-Tomaszczyk, B. (ed.). *PALC 2009: Practical Applications in Language and Computers*. Frankfurt am Main: Peter Lang, 529–546.

Sándorfia, M. (1996) Dance-relevant definitions in John Florio's 1611 Italian-English dictionary. *Letter of Dance*. Vol. 4. [Online]. Available at http://www.pbm.com/~lindahl/lod/vol4/florio.html. [Accessed on 18 August 2020].

Savov, S.A. (2014) Dance narratology (sight, sound, motion and emotion). *Proceedings of the World Congress of the International Association for Semiotic Studies*. [Online]. Available at https://iass-ais.org/proceedings2014/view_lesson.php?id=95. [Accessed on 15 August 2020].

Schäfer, J. (1989) *Early Modern English Lexicography*. 2 vols. Oxford: Clarendon Press.

Sorell, W. (1957) Shakespeare and the dance. *Shakespeare Quarterly* 8(3), 367–384.

Sorell, W. (1967) *The Dance through the Ages*. New York: Madison Square Press.

Stuffer, D. (2016) Visualizing English print: Textual analysis of the printed record. [Online]. Available at https://graphics.cs.wisc.edu/WP/vep/. [Accessed on 20 August 2020].

Swain, J.P. (2013) *Historical Dictionary of Baroque Music*. Lanham, MD/Toronto/Plymouth: The Scarecrow Press.

Teubert, W. (2001) Corpus linguistics and lexicography. *International Journal of Corpus Linguistics* 6(1), 125–153.

Wiesner, S.L. (2011) *Framing Dance Writing: A Corpus Linguistics Approach*. Riga: Lambert Academic Publishing.

Wiesner, S.L. (2019) Establishing the specialist language of dance. In: Fitzsimmons, P., Charalambous, Z. and Wiesner, S. (eds.) *Spectrums and Spaces of Writing*. Oxford: Brill, 13–21.

Williams, G. (2003) From meaning to words and back: Corpus linguistics and specialised lexicography. *ASp: La revue de GERAS* 39–40, 91–106.

Winerock, E.F. (2016) Dancing in Twelfth Night: Courtly versus carnal entertainments. [Online]. Available at https://shakespeareandance.com/articles/dancing-in-twelfth-night-courtly-versus-carnal-entertainments. [Accessed on 31 October 2020].

Wright, F.A. (1916) The technical vocabulary of dance and song. *The Classical Review* 30(1): 9–10.

Yuzurihara, A. (2005) The concept of 'temps' as ballet terminology. *Bigaku* 56(2), 55–68.

Yuzurihara, A. (2013) The construction of classical dance vocabulary in the light of the principle of variation: a comparison with the compositional techniques of contemporary dance. *Comparative Theater Review* 12(1), 133–145.

Zampolli, A. (1983) Lexicological and lexicographical activities at the Istituto di Linguistica Computazionale. In: Zampolli, A. and Cappelli, A. (eds.) *The Possibilities and Limits of the Computer in Producing and Publishing Dictionaries*. Pisa: Giardini, 237–278.

Zorn, F.A. (1905) *Grammar of the Art of Dancing* (1887), ed. by A.J. Sheafe. Boston, MA: Heintzemann Press.

Part III

Analysis

The lexicon of dance in early modern English plays: An annotated glossary

A

Almain/Allemand(e)

Allemande[1] (Almand, Almayne, Almain, etc.) (Fr.) The name of two distinct types of composition, both probably of German origin.

1. Dance, usually in 4/4, but sometimes in duple time, much used by 17th- and early 18th-century composers as the first movement of the suite, or the first after a prelude. It is serious in character but not heavy, and of moderate speed: it is simple binary form.
2. Peasant dance still in use in parts of Germany and Switzerland. It is in tripletime and of waltz-like character. Occasionally composers have called a composition of this type a *Deutscher Tanz* (plural *Deutsche Tänze*), or simply *Deutsch* (plural *Deutsche*).

Concise Oxford Dictionary of Music

Enter Nine Knights in armour, treading a warlike ***almain****, by drum and fife; and then they having marched forth again, Venus speaks.*

(G. Peele, *The Arraignment of Paris*, 1581–84, 2.2)[2]

Introductio in Actum tertium.

Before this act the Hobaies sounded a lofty ***Almain****, and Cupid ushereth after him, Guiszard and Gismund hand in hand. Iulio and Lucrece, Renuchio and another maiden of honor. The measures trod, Gismunda gives a cane into Guiszard's hand, and they are all led forth again by Cupid. Et sequitur.*

(R. Wilmot, *Tancred and Gismund*, 1591, 3)

ALEXANDER: Please it your highness to dance with your bride?
EDWARD: Alas! I cannot dance your German dances.
BOHEMIA: I do beseech your highness, mock us too!
We Germans have no changes in our dances
An **Almain** and an upspring that is all,
So dance the princes, burghers, and the bowrs.
BRANDENBURG: So danc'd our ancestors for thousand years.
EDWARD: It is a sign, the Dutch are not new-fangled.
I'll follow in the measure; marshal, lead!

ALEXANDER and MENZ have the fore-dance with each of them a glass of wine in their hands; then EDWARD and HEDEWICK,

> *PALSGRAVE and EMPRESS, and two other couples, after Drum and Trumpet.*
>
> (G. Chapman and/or G. Peele, *Alphonsus*, 1594, 3.1)

Above are the only three occurrences of the lemma *Almain* – with possible spelling variations – that #Lancsbox found in the entire EMEP corpus. Even looking at EEBO-TCP as a reference corpus, Almain and Allemand(e) occur respectively some 478 and the 24 times in early modern English texts from the 1480s to the 1690s. In the majority of the cases, Almain/Allemand(e) is used as an adjective to indicate the geographical origin or ethnicity of a subject/object (e.g., Almain footmen, Almain emperor, Allemand president, etc.); this connotation is somehow adopted also by George Chapman/George Peele[3] in *Alphonsus*, as will be seen later on.

Therefore, although the hits of ALMAIN[4] are quantitatively of little significance, their concordances and left/right context provide important information about this dance. For instance, the Almain danced in Peele's[5] first play *The Arraignment of Paris* is defined as warlike, meaning a belligerent-like performance accompanied by drums and fifes. Similarities between dancing and fighting have been recently considered by scholars (see, among others, Anglo 2007; Pugliese 2015; Shaw 2019), who tend to underline the close physical connection between dance steps and war/military techniques. As Payne observes, "[t]he almain was often referred to in a way which implies military style and a marching tempo" (2003, 127). For this reason, the Almain performed by the nine knights in *The Arraignment of Paris* – actually the oldest occurrence of the lexeme itself according to the *OED* – appears as a grave and solemn courtly dance that could also be performed by uni-sex dancers.

Written and performed by gentlemen of the Inner Temple at Greenwich before Queen Elizabeth in 1566–68, *Gismond of Salerne*[6] was ampliated and published for the first time in 1591 by Robert Wilmot, who entitled it *Tancred and Gismund*. Among the additions to the original play, Wilmot inserted some dumb shows to introduce each act, except act 1. Before act 3 begins, an Almain is performed on stage by couples, this being the first performance ever attested of a dumb show danced by characters "without any symbolic connotation" (Mehl 2011, 55). As with many early modern dances inserted within stage actions, even in this case the function of the Almain is to "set the scene for the conversation which follows" (Mehl 2011, 55), thus giving the couples, especially the protagonists

Guiszard and Gismund, the chance to come closer without attracting too much attention. Left concordances tell us that the Almain performed is defined 'lofty', "of a noble or elevated nature" (*OED*, adj. 1.1). The presence of the adjective 'lofty' clearly identifies the Almain as a solemn courtly dance, since it is performed in Tancred's Court. *Tancred and Gismund* is set in Naples, with Boccaccio's novella *Tancredi e Ghismunda* its main source. The Almain was quite widespread in Renaissance Italian courts, as Payne has argued (2003, 43) and as evidenced by Italian terpsichorean manuals.[7] Moreover, as *Gismond of Salerne* first and *Tancred and Gismund* later were staged by gentlemen of the Inner Temple, it was absolutely normal that they performed a choreography they knew quite well, given that four out of the eight Old Measures listed in the manuscripts at the Inns of Court were Almains, as described in Part I.

A more precise connotation of the Almain is given by Chapman and/or Peele in *The Tragedy of Alphonsus, Emperor of Germany*, set, as the title reads, in thirteenth-century Germany, where the Almain is thought to have originated, even judging by its name.[8] This dance is once again performed at Court, hence its solemn character. Prince Edward of England (then Edward I Plantagenet), son of King Henry III, is visiting Alphonsus's Court when he falls in love with Hedewick, daughter of the Duke of Saxon, one of the seven electors of the German Empire. A marriage is arranged and during the celebrations Edward is invited by Alphonsus's page, Alexander, to dance with Hedewick. When Edward admits he cannot dance German dances, the King of Bohemia and the Marquess of Brandenburg reassure him that German dances are few and easy to dance, and Bohemia mentions both the Almain and the Upspring,[9] the latter more a general term to define any kind of wild dance than an proper choreography per se (see Elze 1867, 144). Therefore, in *Alphonsus*, the Almain is associated both with the Court – as in *Tancred and Gismund* – and with Germany, a ball which, together with more folk/popular dances like the Upsprings, is danced by everyone, from princes to peasants.

B
Bergomask

> **Bergamasca** (also bergomask, bergamasco, bergomasque) An old Italian peasant dance, from the area around Bergamo in Lombardy. Believed to be a fast circular dance in duple time for men and women, although there are no extant records or notations. It originated in the mid-15th century.
>
> *Oxford Dictionary of Dance*

BOTTOM: Will it please you to see the epilogue, or to hear a **Bergomask** dance between two of our company?

THESEUS: No epilogue, I pray you; for your play needs no excuse. Never excuse; for when the players are all dead, there needs none to be blamed. Marry, if he that writ it had played Pyramus and hanged himself in Thisbe's garter, it would have been a fine tragedy: and so it is, truly; and very notably discharged. But come, your **Bergomask**. Let your epilogue alone.

Bottom and Flute dance a Bergomask, then exeunt.

(W. Shakespeare, *A Midsummer Night's Dream*, 1596, 5.1)

The Bergomask performed by the mechanicals in Shakespeare's *A Midsummer Night's Dream* is a hapax. No other mentions are reported in the wordlist available on the VEP Expanded Corpus 1660 webpage (https://raw.githubusercontent.com/uwgraphics/VEP-corpora/master/core_drama_1660_v2-1grams.csv), nor do the #Lancsbox Words and KWIC functions identify any other occurrence when searching the string berg*mas*.[10] Even the results obtained by looking for the same string on EEBO are not satisfactory, although some 123 hits appear as adjectives referring to the Italian city of Bergamo and its dialect which, apparently, did not enjoy widespread appreciation in early modern England, judging from the collocational patterning of EEBO's occurrences. To mention only a couple of examples that might have interested Shakespeare, in Castiglione's *Il Cortegiano* – rather, in its 1561 English translation by Thomas Hoby – Bergamo's dialect is considered the worst in Italy. Indeed, one reads that "the Bergamask tongue [is] the most barbarous in Italy" and "the silly fellow spoke still his native language the mere bergamask tongue: the worst speech in all Italy". Moreover, in *A Plain and Easy Introduction to Practical Music Set down in Form of a Dialogue* (1597), Thomas Morley[11] comments on the Justiniana, a song from Bergamo whose name derives from a famous prostitute of the city, and defines it as "a wonton and rude kind of music", for it was written "in the Bergamask language" (see Holland 1998, 250–251).

Moreover, searching for the word 'Bergamo' in the #Lancsbox Words function, it emerges that in the EMEP corpus there are 7 occurrences of this lexeme and its left/right concordances on KWIC and GraphColl show very negative connotations. Indeed, the one hit in Shakespeare's *The Taming of the Shrew* (1591) does not connotate the city and its inhabitants positively. In the dialogue between Tranio (Lucentio's servant disguised as his master) and Vincentio (Lucentio's real father coming from Pisa to Padua) in 5.1, the manservant, still trying to convince the Pisan rich man that he is his son Lucentio, affirms that he is able to maintain his "pearl and gold" thanks to his father. Vincentio, tired of being mocked by Tranio and the others, insults Lucentio's servant, declaring scornfully that his father "is a sailmaker in Bergamo", so he cannot afford neither pearls nor gold.

Another negative connotation of Bergamo is present in Thomas Dekker and Thomas Middleton's *The Honest Whore, Part 1* (1604). In 4.4, when Infelice is sent to Bergamo by his father, the Duke of Milan Gaspero Trebazzi, her lover Hippolito is desperate because he believes she has died. However, the doctor reveals to him the Duke's plan and Hippolito states that for her he will ride to "Bergamo, in the low countries of black hell" – even though here Bergamo is simply the place where Infelice was confined, not a devilish inferno.

With this context in mind, the dance performed by the rude mechanicals in *A Midsummer Night's Dream* cannot be considered a solemn courtly performance. Nevertheless, on the one hand, the Bergomask is "raucous [and] rustic" (Hoskins 2005, 19), hence associated with folk environments not that appropriate to a Duke's Court – although, as Hiscock notes, "the early modern [...] courts were not wholly unfamiliar with the antics of Puck and the clowning mechanicals" (2018, 54). According to Skiles Howard, the fact that the Bergomask is the direct object of the verb 'to hear' "locates it within a popular, oral tradition" (1993, 337). On the other hand, it is a vivid and energetic performance particularly suited to celebrating the "aristocratic carnival" (Wiles 2008, 213) of the three marriages which took place in Theseus's Athenian palace. After all, Shakespeare's propensity for not respecting traditional decorum is well-established (see, i.a., McAlindon 1973), as well as *A Midsummer Night's Dream*'s mixing of elevated moments typical of tragedy and purely comic scenes (see, i.a., Hutton 1985).

Walter Sorell describes the Bergomask as "a round dance of couples in which dancers execute little entrechats" (1957, 383), and Peter Holland concludes that it seems an "ordered dance and certainly not a comic one" (1998, 250). Whatever its character, what is certain is that the Bergomask was performed by couples and no role-related gender-related differences seem to be suggested; in fact, quite the opposite. Contrary to the fairies' Carols, which celebrate a harmonious universe re-established thanks to the fact that "patriarchy is restored as Oberon overcomes Titania's rebellion against his wishes" (Lamb 2013, 309), the Bergomask of the artisans is performed presumably by the protagonists of the play-within-the-play, Pyramus and Thisbe, who are deliberately and metatheatrically a male-male couple, not only because women could not act on stage in early modern England, but also because the role of Thisbe is assigned to the reluctant bellows-mender Francis Flute. Therefore, the Bergomask performed by Nick Bottom,[12] who plays Pyramus, and Francis Flute, overtly disguised as Thisbe, does seem to be a gender-neutral dance where the distinction between man and woman disappears, perhaps even recalling the Neoplatonic ideal of the primordial androgynous being – definitely a perfect conclusion to celebrate three marriages.

Branle/Brawl

> **Branle** (Fr., swing, shake; can also be spelled bransle) The term for old French folk dances, frequently accompanied by singing. In modified form they found their way into the entertainments at the court of Louis XIV and hence later became connected with various ballets and pantomimes. In his *L'Orchésographie* (1588), Thoinot Arbeau described more than two dozen kinds of branles. In the 20th century the branle is featured in Balanchine and Stravinsky's *Agon*.
>
> *Oxford Dictionary of Dance*

AURELIA: We will dance, music, we will dance.
GUERRINO: Les quanto (Lady) penses bien, passa regis, or Bianca's **brawl**.
AURELIA: We have forgot the **brawl**.
FERRADO: So soon?'tis wonder.
GUERRINO: Why't is but two singles on the left, two on the right, three double forward, a traverse of six round: do this twice, three singles side, galliard trick of twenty, coranto pace; a figure of eight, three singles broken down, come up, meet two doubles, fall back, and then honor.

AURELIA: O Daedalus! thy maze, I have quite forgot it.
(J. Marston, *The Malcontent*, 1604, 4.2)

LADISLAUS Let the masquers enter: by the preparation,
'Tis a French **brawl**, an apish imitation
Of what you really perform in battle.
And Pallas, bound up in a little volume,
Apollo, with his lute, attending on her,
serve for the induction.
Enter two boys, dressed as Apollo with his lute and Pallas: A Dance; after which a Song by Pallas, in praise of the victorious soldiers.
(P. Massinger, *The Picture*, 1629–30, 2.2)

When dealing with the Brawl, a problem arises due to the word's polysemy in English. Indeed, the primary meaning of the noun and the verb '(to) brawl', according to the *OED*, is respectively "A rough or noisy fight or quarrel" (n. 1) and "Fight or quarrel in a rough or noisy way" (v. 1). The dialogue between the Spanish knight Armando and his page Mote in Shakespeare's *Love's Labour's Lost* (1594), albeit not interesting for an analysis of the connotations of this dance per se, offers a perfect example of the comic effect that a misunderstanding due to polysemy may have:

MOTE: Master, will you win your love with a French **brawl**?
ARMANDO: How meanest thou? **Brawling** in French? (3.1)

Therefore, attention must be paid to the occurrences of the string 'brawl*' (i.e., brawl, brawle, brawls, brawles, etc.; 124 in total in the EMEP corpus analysed here) when it is searched in the KWIC tool. Nevertheless, the GraphColl function and the lexeme's collocational patterning helped to distinguish 8 occurrences of the word which refer exclusively and undoubtedly to the dance of French origin. In fact, when brawl is reported by playwrights with its primary (negative) meaning, in the corpus it collocates with such nouns as 'broil', 'discontent', 'fury', 'hate', 'jangling', 'quarrel', 'scold', 'wrangle', etc., verbs like 'avoid', 'begin', 'close', 'fling', 'fall into', 'end', 'swear', 'swagger', etc., and adjectives like 'luckless', among others. On the contrary, the occurrences of Brawl as a dance are signalled by the two main collocations 'dance' and 'French', this latter adjective leaving no doubt about the origin of this choreography.

In the early 1600s, it was John Marston, member of the Middle Temple since 1592, who mentioned the Brawl as a dance of French origin. Although in *Jack Drum's Entertainment* (1600–1, 5.1), the Brawl is not performed but only its music played – exactly as in John Day's *Humour Out of Breath* (1607–8, 2.2) – in *The Malcontent*, Marston reports one of the fewest and hence most important directions to follow to perform a dance in early modern English plays. The Genoese courtier Guerrino, perfect embodiment of the dance-skilled gentleman praised by Renaissance conduct books and terpsichorean manuals, repeats the sequence of steps necessary to perform Bianca's Brawl[13] to the Duchess of Genoa, Aurelia, who has completely forgotten it but is now eager to dance it. As Hoskins explains, the Brawl is a "rowdy and boisterous" dance, this aspect reflecting the scene where it is set as well as the play itself. According to Clark (1983, 85ff.), the dance is perfectly inserted within a play full of references to sexual intercourses and cuckoldry, and the chaotic Brawl suits the chaotic personality of Aurelia, who cheats on her husband Pietro and contemplates killing him. In such a misogynistic context, where cuckoldry is produced by women, not by God, as the malcontent Malevole asserts, Aurelia's eagerness to dance a Brawl mirrors the woman's natural tendency to cheat on her husband, the Duke of Genoa.

Given that it is a French dance, it is not surprising that the Brawl and the imagery associated with it recur in the Caroline era, when Queen Henrietta Maria[14] brings a new wave of French (and Italian) influences to the English Court. In Philip Massinger's *The Picture* a masque is performed at Ladislaus's Hungarian Court to celebrate the victory against the Turks. During this masque, a Brawl is danced and sung by Pallas, goddess of war, and accompanied on the lute by Apollo, god of music, with the purpose of imitating the battle won by general Ferdinand thanks to Matthias's astonishing strength and bravery. Of course, the Brawl has a double function here; on the one hand, its lively rhythm is well suited to the excitement and confusion of a battlefield. On the other hand, it is once again associated with corruption and cuckoldry: Ladislaus's Court is a sink of lust where the two courtiers Ubaldo and Ricardo pursue any woman they find attractive – even Matthias's faithful wife Sophia – and where Queen Honoria tries by any means possible to instigate a sexual affair with Matthias. Ladislaus is a weak king, victim of his wife's wild desires, as the counsellor Eubulus notices.

No positive connotations of the French Brawl are given by James Shirley either in *The Example* (1634–37, 1.1) or in *The Lady of Pleasure* (1635–37, 3.2), two of his most appreciated comedies. In the first case, the French Brawl is associated with gambling by the paranoiac Sir Solitary Plot, convinced that both his wife and niece Bellamia stay out at night playing, dancing, fornicating, and plotting against him. In *The Lady of Pleasure*, on the other hand, the ability to dance the Brawl is used by the sixteen-year-old widow Celestina as an opportunity to rail against Kickshaw and Littleworth, "the ignobler beasts" who pursue Lady Aretina Bornwell, the lady of pleasure of the title. The two suitors, says Celestina in front of an angry Lord Bornwell, are only capable of dancing the Brawl – the boisterous dance being an undisputed metaphor for sexual intercourse – and nothing else.

As Winerock pointed out (2011, 455), in Italian and French Renaissance manuals, the Brawl/Branle was one of those dances "that required women to ask men to dance", since

> the leader of the dance chose a partner of the opposite sex, danced with that person, handed over the object, and then retired. The new possessor of the object would choose another partner, and the dance would continue in this pattern. Thus, every other repetition of the pattern required a woman to ask a man to be her partner.

What I would add in this context – further bolstering Winerock's remarks – is that given the 'promiscuous' nature of the Brawl (as Puritans would say), this dance is often associated with extramarital affairs and deviant sexual behaviour charges in early modern English plays.

C

Canary

Canary A jaunty toe-tapping dance in 3/4 or 6/8 time. It is similar to the jig and the gigue. It probably originated in the Canary Islands; hence its name. Its first recorded appearance in European dance history was in Fabritio Caroso's dance manual *Il ballarino* (1581). The first musical examples of canaries are found in the harpsichord suites of Couperin and de Chambonnières. The dance is mentioned in Shakespeare's *All's Well That Ends Well*: "Make you dance Canary with spritely fire and motion".

Oxford Dictionary of Dance

BELLAFRONT: Acquaintance, where supp'd you last night?
CASTRUCHIO: At a place, sweet acquaintance, where your health danc'd the **canaries**, i'faith; you should ha' been there.
BELLAFRONT: Ay, there among your punks. Marry, fah, hang'em! Scorn't!

(T. Dekker and T. Middleton, *The Honest Whore, Part 1*, 1604, 2.1)

LAFEU I have seen a medicine
That's able to breathe life into a stone,
Quicken a rock, and make you dance **canary**
With spritely fire and motion; whose simple touch,
Is powerful to araise King Pepin, nay,
To give great Charlemain a pen in's hand,
And write to her a love-line.

(W. Shakespeare, *All's Well That Ends Well*, 1605, 2.1)

COOK I'll make yee pigs speak French at table, and a fat swan
Come sculing out of England with a challenge.
I'll make yee a dish of calves feet dance the **Canaries**,
And a consort of cram'd capons fiddle to'em.
A calf's head speak an oracle, and a dozen of larks
Rise from the dish, and sing all supper time;
Tis nothing boys, I have fram'd a fortification,
Out of rye past, which is impregnable,
And against that for two long hours together,
Two dozen of maribones shall play continually.
For fish I'll make ye a standing lake of white-broth,
And pikes come ploughing up the plumbs before 'em

Arion on a dolphin playing lachrimae,
And brave King herring with his oil and onion
Crowned with a lemon pill, his way prepar'd
with his strong guard of pilchers.
(J. Fletcher et al., *The Bloody Brothers*, 1612–24, 2.2)[15]

SOTO To a banquet there must be wine. Fortune's a scurvy whore, if she makes not my head sound like a rattle and my heels dance the **canaries**.
(J. Ford et al., *The Spanish Gipsy*,[16] 1623, 4.2)

Even in the case of the Canary, it was necessary to deal with its polysemy before proceeding to the analysis of its connotations in the EMEP corpus. In fact, the lemma *canary* has different meanings – at least four – according to the *OED*. It can indicate 1) "A mainly African finch with a melodious song, typically having yellowish-green plumage. One kind is popular as a cage bird and has been bred in a variety of colours, especially bright yellow" (n. 1), 2) "A sweet wine from the Canary Islands, similar to Madeira" (n. 3), 3) the Canary Islands themselves, or 4) a dance of Spanish origin. No reference to canary yellow ("A bright yellow colour resembling the plumage of a canary", *OED*, n. 2) appears in the EMEP corpus. Here again, the #Lancsbox GraphColl function enabled me to divide the 80 occurrences of the string 'canar*' into four different groups according to the various collocational patternings of the node sought. When Canary meant the Mediterranean Spanish isles, the string collocated with prepositions such as 'in the' or 'to the', and nouns such as 'land' or 'isle'. When it indicated the yellow bird, such words as 'bird' or other animal species formed its left/right context, and lastly the meaning of wine collocated with nouns, adjectives, verbs, and quantifiers such as 'a cup of', 'a pint of', 'a quart of', 'glass', 'rich', 'sack', and 'wine'.

The four occurrences of the Canary as a dance in the EMEP corpus belong to comedies or comic scenes within tragedies and tragicomedies, thus highlighting the vivid and lively character of this dance of love "in which dancers began to pantomime innuendos of lovemaking" (Hoskins 2005, 29–30). Both men and women performed some movements alone, while the other partner simply stood and looked on, and the steps were so difficult and energetic that Arbeau compares the Canary to the dances of the "savages", full of erotic charge. In Dekker and Middleton's *The Honest Whore, Part 1*, Bellafront, the honest whore of the title, and the gallant Castruchio, one of her regular customers, speak about everything and anything. Their dialogue, together with that of the other two gallants Fluello and Pioratto, is full of double entendre, given the setting, a brothel, and the characters involved. Castruchio mentions the Canary in the idiom "your health danc'd the canaries" to indicate that he had great fun and probably sex as well. Although one cannot affirm that Castruchio actually danced a Canary in the place where he had supper the previous night, the imagery that this dance recalls is certainly not that associated with solemn courtly dances. By using the Canary as a metaphor, the gallant might wish to pass himself off as a great and tireless lover.

Lord Lafeu's medicine in Shakespeare's *All's Well* can make the King of France "dance canary". Lafeu uses the imagery associated with the Canary hyperbolically, to convince the King to take a medicine brought to the royal palace by a woman, Helen. The king seems to be in favour of such a miraculous drug, as he knew Helen's late father and was familiar with his excellent skills as a physician. In this context, the Canary implies no sexual intercourse, but is only a synonym for energy and vitality. What is worth noting here is that the Canary was known both at Court and outside of it – as shown by the scene from *The Honest Whore, Part 1*.

Again, a hyperbolic connotation of the Canary is present in Fletcher *et al.*'s *Rollo* when the Court cook lists his culinary abilities. In his overstated speech, he affirms that he can prepare a dish with calves' feet that can dance the Canary. If not taken literally, the cook's hyperbolic assertion underlines his supposed astonishing skills. Indeed, the Canary, being a fast-paced dance, was known for its incredibly rapid alternation of heel and toe stamping movements of the feet – the cook implying that, when cooked by him, the feet of any calf could do incredible things. This interpretation is supported by the fact that, according to Hoskins, the Canary "was considered so difficult that only the most practiced dancers attempted it" (2005, 30).

Lastly, the Canary's quickness of step is also used metaphorically in John Ford, Dekker, Middleton, and William Rowley's *The Spanish Gypsy*,[17] certainly the work where the mention of a Canary is more appropriate than elsewhere, given the origin of the dance and the setting of the play in Spain. While preparing the play-within-the-play written by don Fernando, the gypsies make some decisions about the scenic design and the assignment of the roles. Sancho and his servant Soto play the servants Hialdo and Lollio, but they want a banquet full of food and wine. Aside from the pun on the polysemy of the lexeme CANARY which also indicates a wine, as noted by Taylor and Lavagnino (2007, 1754), Soto refers again to the rapid movements of the feet – this time, metonymically, the heels – in executing the steps of the Canary.

Thus when not associated with sexual intercourse, the Canary metaphorically refers to something incredibly vivacious and energetic. Given Hoskins's above-mentioned statement regarding the extreme complexity of dancing the Canary, it is no coincidence that no early modern play, as far as can be ascertained from the analysis presented here, actually staged it, with playwrights limiting themselves to using the imagery associated with it.

Cinquepace/Sinkapace

> **Cinquepace** (sinkapace), one of a family of couple dances imported from France and Italy, including the galliard, tourdion, and la volta; also and alternative name for the galliard. The unit of five steps implied by the name fits to a six-beat bar: four springing kicks on the first four beats, with a jump through the fifth beat, timed so as to land on the sixth beat.
>
> *Oxford Companion to Shakespeare*

> BEATRICE If the Prince be too important, tell him there is measure in everything, and so dance out the answer. For hear me, Hero, wooing, wedding, and repenting is as a Scotch jig, a measure, and a **cinquepace**. The first suit is hot and hasty like a Scotch jig, and full as fantastical; the wedding, mannerly modest as a measure, full of state and ancientry; and then comes repentance, and with his bad legs falls into the **cinquepace** faster and faster till he sink into his grave.
>
> (W. Shakespeare, *Much Ado About Nothing*, 1598, 2.1)

> FREDERICK I think my father thinks I am an ass:
> Cannot I lead a lady by the arm,
> Hold off my hat, and dance my **Cinquepace**.
>
> (J. Cumber, *The Two Merry Milkmaids*, 1619, 2.2)

FRISK: 'Tis but dis in de beginning, one, two, tree, four, five, the **cinquepace**; *allez*, *monsieur*! Stand upright, ah! Begar.

LORD RAINBOW: Let him set you into the posture.

FRISK: My broder, my lord, know well for de litle kit, de fiddle, and me for de posture of de body; begar de king has no two such subjects; hah! Dere be one foot, two foot, – have you tree foot? Begar you have more den I have den.

BARKER: I shall break his fiddle.

LORD RAINBOW:	Thou art so humorous. [*Barker dances.*]
FRISK:	One, *bien*! Two; – hah, you go fast! You be at Dover, begar, and me be at Greenwish; tree, – toder leg, pshaw!
BARKER:	A pox upon your legs! I'll no more.
	(J. Shirley, *The Ball*, 1632–39, 3.1)

BLOOD:	This leg's not right.
TASTE:	I know it.'Tis my left.
BLOOD:	Carry your toes wider.
TASTE:	Take heed that I foot not you.
BLOOD:	Now do your **cinquepace** cleanly.
TASTE:	My **cinquepace** cleanly! A cook defies it.
	(T. Nabbes, *Microcosmus, a Moral Masque*, 1637, 4)

Little can be said about the Cinquepace from a collocational point of view,[18] given that its collocational patterning does not highlight recurring features, except its pairing with possessives such as 'my' or 'your' (see Cumber's *Milkmaids* and Nabbes's *Microcosmus*). Moreover, given the name of this dance and its execution, some of the hits collocate with numbers and calculations (e.g., Shirley's *The Ball*, but also William Cavendish – and Shirley's? – *The Country Captain*, 1640), as a pun.

For the girl born when "a star danced", "there is measure in everything". In *Much Ado About Nothing* Shakespeare's Beatrice mentions the Cinquepace to her cousin Hero as one of the three dances comprising the cycle of love. If wooing is as fast as a Scottish Jig, the wedding is a solemn and placid Measure, while repenting is again as fast as a Cinquepace, until the love ends. Therefore, this dance is used here for the speed of the execution of its steps. Unlike the Scottish Jig, which, being a more folkloric dance, represents the freshness and carefreeness of the moment when a couple falls in love, the courtly Cinquepace is represented here as a rapid and inevitable ("faster and faster") succession of steps that take love to the grave.

A different connotation of the Cinquepace emerges from the context in which it is mentioned in Frederick's complaint against his father Lodowicke in John Cumber's *The Two Merry Milkmaids, or the Best Words Wear the Garland*. In the character's words, the Cinquepace is one of the gallant actions that his parent accuses him of not being able to perform towards a woman. In other words, as the Cinquepace is a proper courtly choreography, Frederick is portrayed as an inadequate courtier and gentleman by his father.

There is probably no other early modern play devoted to dance as James Shirley's *The Ball*, whose title anticipates the centrality of the terpsichorean art in the plot – and subplot – of the comedy. Indeed, the play depicts members of the upper class organising a fashionable dancing event. Of course an exuberant French dancing master, Monsieur Frisk,[19] is brought in and given the task of teaching the various characters to dance. When Lord Rainbow and Frank Barker have a lesson with Monsieur Frisk, the dancing master accepts Lord Rainbow's proposal to include the cynic Barker as a pupil and begins correcting both his steps and posture while performing the Cinquepace. Barker, whose language is definitely not polite, is completely uncomfortable with courtly dances – this scene being a prelude to his embarrassing dance as a satyr at the request of Honoria – and Lord Rainbow is the only one who laughs at this apish scene. Monsieur Frisk is intransigent and the fast rhythm of the Cinquepace reinforces the fast pace of the entire scene – so much so that when Frisk reproaches Barker for his slowness in executing the dance's basic steps, Barker stops dancing. Therefore, once again it is the quickness of the choreography that is highlighted in this scene.

Exactly as in the embedded masque in Massinger's *The Picture*, Nabbes's *Microcosmus* is one of the few masques where dance names are actually reported, and, I would add, also one of the few plays where professional dancers or dancing masters are actual characters (as in *The Ball*). *Microcosmus*, defined as a moral masque, is basically a Caroline morality. In the scene quoted above, the dancer Blood is teaching Taste, one of the five senses, to dance the Cinquepace. Nevertheless, Blood, Taste, and the other three complexions (the physician Phlegme, the fencer Choler, and the musician Melancholy) are drunk, and each has a bottle of wine in the hand; thus the dance lesson predictably fails – even because of their puns. The Cinquepace, as Blood states, is a complex courtly dance requiring clean movements, but Taste bungles again with the polysemy of the adverb 'cleanly'.

Coranto/Courante

Courante (Fr.), corrente (It.), coranto, corant. Running. French dance, at height of popularity in 17th century, which spread to Italy. The music based on it falls into 2 classifications. (*a*) Italian variety, in a rapid tempo and in simple triple time. (*b*) French variety, similar to the above, but with a mixture of simple triple and compound duple rhythms, the latter pertaining especially to the end of each of the 2 sections. Occasionally in Bach's kbd.

examples the conflicting rhythms are found together, one in each hand. In classical suite the courante followed the allemande [...]. Occasionally it was, in turn, followed by 'Doubles', i.e. variations on itself.

Concise Oxford Dictionary of Music

BRETAGNE They bid us to the English dancing-schools,
And teach lavoltas high and swift **corantos**;
Saying our grace is only in our heels,
And that we are most lofty runaways.

(W. Shakespeare, *Henry V*, 1599, 3.6)

SECOND MAID Come on, old grummelseeds. What, must we stand thrumming of caps all day, waiting on your grave ignorance? By the faith of my body, either let your daughter dance with us, or I'll make your old bones rattle in your skin; I'll lead you a **Coranto** I'faith.

(P. Hausted, *The Rival Friends*, 1631, 4.9)

LITTLEWORTH: Is there any man that will cast away his limbs upon her?
KICKSHAW: You do not sing so well as I imagined, nor dance; you reel in your **coranto**, and pinch your petticoat too hard: you've no good ear to the music, and incline too much one shoulder, as you were dancing on the rope, and falling. You speak abominable French, and make a courtesy like a dairy maid – Not mad?

(J. Shirley, *The Lady of Pleasure*, 1635–37, 3.2)

CUTTER: O divine Tabytha! Here come the fiddlers too. Strike up, you rogues.
TABYTHA: What? must we dance now? is not that the fashion? I could have danc'd the **Coranto** when I was a girl. The **Coranto**'s a curious dance.

(A. Cowley, *The Guardian*, 1641, 5.6)

The two adjectives which collocate with the Coranto are 'swift' and 'curious'. Therefore, on the one hand, the dance is associated with the rapidity of its steps; on the other, the adjective 'curious' meant both "sophisticated" and "requiring care and art" (*Oxford Dictionary of English Etymology*), and Cawdrey used it as synonym for 'accurate' and 'exquisite' in his *A Table Alphabetical* (1604).

According to Helen Ostovich (1994), both cross-national and gender issues are contemplated in the Duke of Bretagne's lines from Shakespeare's *Henry V*. The contrast between French – caricatured "as inept villains and lustful clowns [...] in several plays on stage between 1598–1600" (154) – and English lords and soldiers is of great importance in the play. In the lines quoted above, Bretagne is complaining about French women who prefer "the lust of English youth" (as the Dauphin had affirmed few lines earlier) to their male compatriots. According to French women, Frenchmen only learnt to dance the La Volta and Coranto in English dancing schools, punning at the fact that they are able only to turn (La Volta) and run away (Coranto), as Craik noticed (1995, 228). In this case, the adjective 'swift', which qualifies the Coranto as a quick dance, acquires a negative connotation in Bretagne's utterance, as if it were synonymous with 'hurried', 'hasty', or 'rushed'.

The same (negative) connotation of fast dance that the Coranto has appears to be present also in Peter Hausted's *The Rival Friends*, when one of the two maids almost threatens to make the shepherd Stipes's bones rattle in his skin with a Coranto, should he not let his daughter Merda dance with them on the green. Clearly, Lively's maid's threat works, as Stipes allows Merda to dance "for half an hour, no more". Therefore, both Shakespeare's and Hausted's mentions of the Coranto seem to pinpoint the dance's negative connotation of a performance so fast that it may appear to be too rapid.

Once again, gender issues are raised by Littleworth and Kickshaw's dialogue with Celestina in Shirley's *The Lady of Pleasure*. After the widow's nth attack against Lady Bornwell's two foppish suitors, Littleworth asks his friend whether any man will ever manage to cast away his limbs upon her – metaphorically echoing coitus. Kickshaw reminds his friend about his 'little worth' – thus revealing once and for all that Littleworth is a speaking name – which stems from the fact that he cannot sing, nor dance the Coranto properly. In other words, Kickshaw is using the Coranto here as a sexual metaphor to discredit Littleworth's skills as a suitor and lover. At the same time, Littleworth clashes with the image of the good courtier who can dance, a sign of his ability to pursue a woman.

Anti-Puritanical nuances are present in the dance scene in Cowley's *The Guardian*. In fact, Tabytha, educated according to the Puritan ethics, is successfully wooed by Colonel Cutter, who makes her drink, dance, and then go to bed, thus marrying her. In this scene, Tabytha regrets that she didn't learn how to dance the Coranto when she was younger, because the Coranto, like love, is a curious dance requiring care and art. Thus, even in this scene the Coranto may be associated with sexual innuendo. Of course the corruption of a girl educated with Puritan values who begins drinking and dancing immediately before her marriage was unacceptable at the dawn of the Civil War, and so Cowley was forced to seek refuge in Paris before coming back to England and revising his comedy, which was published after the monarchic restoration under the title *The Cutter of Coleman Street* in 1661.

Country Dance

> **Country dance** (Eng.), contredanse (Fr.), contradanza (It.), Kontretanz (Ger.) This type of dance is of British origin. Its various foreign names have come about from a plausible false etymology ('counter-dance' – one in which the performers stand opposite to one another – as distinguished from a round dance). Both Mozart and Beethoven wrote Kontretänze. No. 7 of Beethoven's 12 Kontretänze contains the theme used also in the finale of the Eroica Symphony and other works. The term is generic and covers a whole series of figure dances deriving from the amusements of the English village green. Such dances became popular at the court of Queen Elizabeth I, and during the Commonwealth were systematically described by Playford in his *English Dancing Master*. In the early years of the 19th century the waltz and quadrille drove the country dance out of the English ballroom (with the exception of the popular example known as Sir Roger de Coverley); the folk-dance movement of the 20th century, however, brought it into considerable use again. Scotland has throughout retained a number of its country dances.
>
> *Concise Oxford Dictionary of Music*

MISOGONUS: Trifle not the time then. Say, what shall we have?
What **country dances** do you here daily frequent?
CACURGUS: The vicar of Saint Fools, I am sure, he would crave.
To that dance, of all other, I see he is bent.
SIR JOHN: Faith, no. I had rather have shaking o'th' sheets or sundry flings;
Or catching of quails or what fair Melissa will.
MELISSA: The fool, I see by him, is given wholly to scorning.
ORGELUS: Priest, keep your cinquepace and foot it o'th' best sort.
They dance.
(A. Rudd, *Misogonus*, 1577, 2.4)

CLEM I am so tir'd with dancing with these same black she-chimney sweepers that I can scarce set the best leg forward; they have so tir'd me with their moriscos, and I have so tickled them with our **country dances**, Sellenger's round and Tom Tiler. We have so fiddled it!

(T. Heywood, *The Fair Maid of the West, Part 2*, 1631, 2.1)

FRISK: Sorre! Tat is too much *par ma foy*, I kiss tat white hand, give me one two tree buffets. – *Allez, allez*; look up your countenance, your English man spoil you, he no teach you looke up, pishaw! Carry your body in the swimming fashion, and – *allez, mademoiselle*, ha, ha, ha! So, *fort bon*! Excellent, begar. [*They dance*]

LUCINA: Nay a **country dance**. Scutilla, you are idle, you know we must be at the ball anon; come.

FRISK: Where is the ball this night?

LUCINA: At my Lord Rainbow's.

FRISK: Oh he dance finely, begar, he deserve the ball of de world; fine, fine gentleman! your oder men dance lop, lop with de lame leg as they want crushes, begore, and look for l'*argent* in the ground pshaw! [*They dance a new* ***Country Dance***] – Hah, hah, *fort bon*!

(J. Shirley, *The Ball*, 1632–39, 2.3)

COURT-WIT: Here, Madam, may you see the madman's revels?

SWAIN-WIT: And after that the doctor's tragicomedy.

FERDINAND: Are not your windpipes tuned yet? Sing a catch! So, now a dance! [*Rising from his chair and cavorting about*] I am all air! A-hey! A-hey! I thank thee, Mercury, that hast lent thy wings unto my feet. Play me my **country dance**. Stand all you by. These lasses and these swains are for my company. *He dances a conceited* ***country dance****, first doing his honours, then as leading forth his lass. He dances both man and woman's actions, as if the dance consisted of two or three couples. At last as offering to kiss his lass, he fancies that they are all vanished, and espies Strangelove.*

(R. Brome, *The Court Beggar*, 1640–41, 4.2)

When examining the Country Dance, a few preliminary considerations must be discussed before analysing its connotations and context in early modern English plays. The images associated nowadays with the Country Dance are of couples wearing cowboy hats, loose shirts, jeans, and boots dancing to folk/country music. This is, however, the American version of the European Country Dance, whose name may derive from the French *contredanse* or Italian *contradanza* and thus has nothing to do with a 'country' environment. In fact, the Renaissance *contredanse/contradanza* owes its name to the two lines of male and female dancers that counterpose each other – although, as the above definition from the *Concise Oxford Dictionary of Music* reads, it might be a false etymology and the English-born Country Dance may actually be associated with its popular origins. Only one occurrence of the multiword Country Dance in the EMEP corpus collocates with words which indisputably associate this dance with a rural environment: the last stage direction in Richard Flecknoe's allegorical fiction – a masque, in fact – *The Marriage of Oceanus and Britannia* (1659), which reads "*two Country swains in all their Country bravery, dancing a Jig, or Country dance*".

In Anthony Rudd's comedy *Misogonus*,[20] one of the earliest Elizabethan "regular" plays, the Country Dance is instead inserted in a chaotic scene where Misogonus, his family servant Cacurgus, the shameless priest Sir John, the courtesan Melissa, and Misogonus's attendant Orgulus are shown in their element: spending their nights out drinking, gambling, and dancing. This seems a pejorative connotation of the Country Dance, which somehow links it to popular, lower-class environments. This consideration is reinforced by the fact that the comedy is set in Italy – and this aspect also rules out the possibility that it is an English Country Dance – which is often considered "the academy of man-slaughter, the sporting place of murder, the apothecary-shop of poison for all nations" which had given birth to "false hearted Machiavellians", as Thomas Nashe famously had it in his picaresque novel *The Unfortunate Traveller* (1594).[21]

Another important aspect to be considered is that the Country Dance reached its peak of popularity in the period immediately following the one I am considering here, immediately after Playford's *The English Dancing Master* and its implicit celebration of this dance as the national emblem of an English autochthonous terpsichorean tradition. Nevertheless, in Heywood's *The Fair Maid of the West, Part 2*, the Country Dance is already depicted as the distinctive English feature, in contrast with the multicultural and multilingual settings of Heywood's plot. While trying to escape the Moroccan court of Mullisheg, the English apprentice clown Clem admits being tired of dancing Moriscos with black women and would like to dance such English Country Dances as Sellenger's Round (also a choreography Playford names on list of Country Dances) and Tom Tiler (a Country Dance that

Heywood mentions in his *A Woman Killed with Kindness*, 1603–7, 1.2). In these lines, the Country Dance is contrasted with the Moresca, a dance of Arab origin considered grotesque. Moreover, although Elizabeth Spiller warns us about the risks of "retrofit[ting] our contemporary conceptions of race onto early modern understandings of it" (2011, 3),[22] it is undeniable that Clem's xenophobic considerations are aimed at putting Moroccans – metonymically the Arabs – in a bad light. In this case, the grotesque and savage Moresca/Morisco danced with metaphorical she-chimney sweepers becomes symbolic of a rude, uncultivated society, one that Clem contrasts to the lively and orderly English Country Dances.

The other two occurrences of the Country Dance in the EMEP corpus highlight a different connotation of this choreography, both at the concordance and pragmatic levels. Lucina's impolite utterance "Scutilla, you are idle" in Shirley's *The Ball* underlines the fact that the rich widow's attendant is incapable of sustaining the fast rhythm of the Country Dance which Monsieur Frisk, the exuberant dancing master, has her perform. Likewise, Ferdinand's Country Dance in Richard Brome's *The Court Beggar* is pre-modified by the adjective 'conceited', which, once again, stresses the quickness of the movements of his performance.

Thus in the examples analysed above the Country Dance emerges as a couples dance (Clem would dance it with the Moroccan "black she-chimney sweepers", Lucina and Frisk perform it, and Ferdinand "lead[s] forth his lass") whose primary connotations are its strong English autochthonous[23] character and its rapid execution.

Cushion Dance

> **Cushion Dance** An old dance in which a participant chose a partner by dropping a cushion before him or her, who then knelt on it and bestowed a kiss on the cushion-bearer.
>
> *Concise Oxford Dictionary of Music*

SLIME: I come to dance, not to quarrel. Come, what shall it be? *Rogero*?
JENKIN: *Rogero*? No; we will dance *The Beginning of the World.*
CICELY: I love no dance so well as *John come kiss me now*.
NICHOLAS: I that have ere now deserv'd a cushion, call for the **Cushion-dance**.
BRICKBAT: For my part, I like nothing so well as *Tom Tyler*.
JENKINS: No; we'll have *The Hunting of the Fox*.

SLIME: *The Hay*, *The Hay*! There's nothing like *The Hay*.
NICHOLAS: I have said, I do say, and I will say again–
JENKINS: Every man agree to have it as Nick says!ALL Content.
(T. Heywood, *A Woman Killed with Kindness*, 1603–7, 1.2)

The Cushion Dance is certainly a folk dance which never made its way to the Court (see Winerock 2005). In the lines quoted above from Heywood's most famous domestic tragedy,[24] *A Woman Killed with Kindness*, this dance is inserted in its folk, popular context, since it is mentioned and called for by John Frankford's servants in 1.2, immediately after the musicians have played some tunes for the marriage of John and his wife Anne (the woman killed with kindness of the title) in 1.1. This reference to the Cushion Dance is the only in the EMEP corpus, but it is highly significant. In fact, not only is this dance called for by Nicholas and the other servants, but it is also performed by them at the end of the scene, for all the characters on stage agree to dance it, as Nick wishes. Unfortunately, Heywood provides no details of that performance and, like many of his contemporaries, merely indicates in the stage direction that "*They dance* [...] *after the country fashion*". I would argue, however, that the fact that there is a country fashion at all, and that it is allowed to enter the stage, somehow confirms that early modern theatre is a privileged and comprehensive microcosm where connotations of dancing can be analysed in all their multifaceted nuances, as both courtly and folk performances are actually staged.

However, this scene is also very important because it reports many titles of popular dances, i.e., Rogero, The Beginning of the World, John Come Kiss Me Now, Tom Tiler, The Hunting of the Fox, and The Hay, thus offering a rare source for our understanding of folk dance in the early modern English period, something only Lisa Hopkins seems to have noticed when dealing with this play, in addition to referring to another scene of the play where the macabre dance The Shaking of the Sheets is mentioned (1994, 3). Actually, some of the dances mentioned by Heywood in *A Woman Killed with Kindness* are also present in Playford's *The English Dancing Master*, thus complementing Whitlock's demonstration that Playford named most of his dances after Ben Jonson's and Brome's plays (1999), as seen in Part I.

D
Dance

This lemma certainly requires no definition, nor can examples of co-textual references be provided. And yet, some statistically significant data are worth considering both from a linguistic and cultural point of view. In order to explore quantitatively relevant information about the lexeme DANCE, the 'Whelk' function has been used. This lexeme occurs 1,767 times in the EMEP corpus, in 407 plays out of 666, with a relative frequency of 1.78 per 10k tokens, which is quite high when compared with the relative frequency of the lexeme on EEBO, that is, 0.28.[25] This is a first important datum, as it may indicate that in early modern English plays, mentions of dancing exceed the average occurrences that the terpsichorean lexicon has in other contemporary textual genres, as the reference corpus shows. The lexeme's inflected forms in the corpus are 'dance(s)', 'danced', 'dancer(s)', 'danceth', and 'dancing(s)'. Of course derivative processes intervene; for instance, when the noun 'dance' is created by the verb 'to dance' by conversion. An interesting case to be considered is the two occurrences of the adjective 'dancitive' in Chapman's *Sir Giles Goosecap* (1606, 2.1):

EUGENIA: What means he by this, tro ye? Your Lord is very **dancitive** me thinks.
MONFORD: I, and I could tell you a thing would make your Ladyship very **dancitive**.

In both cases 'dancitive' is pre-modified by the adverb 'very' and uttered while Monford and his niece Eugenia are in fact dancing. Nevertheless, as often happens when dealing with a dance-related lexicon, the adjective has various polysemous connotations. For Eugenia, her uncle Monford is very dancitive because he is speaking and dancing at the same time, but he is also dancitive in the sense that he is ambiguous, elusive about what he wants to tell her – i.e., the poet Clarence's love for her. On the other hand, Monford thinks that the news he has to give her will make her very happy. However, the most interesting aspect of these two occurrences of the adjective 'dancitive' is that they are the only two hits of the word in the EMEP corpus and, even more significantly from a quantitative perspective, in EEBO, as also stated in the *OED* at the lemma *dancitive*. This means that not only might this adjective be a neologism by Chapman created by the same derivative process of sense (n.) → sensitive (adj.), but also that such a morphological and lexical process was not productive or successful.

Looking at the EMEP corpus more closely, it emerges that the play where the lexeme DANCE has the highest number of occurrences is, not surprisingly, Shirley's *The Ball*, with its absolute frequency of 31 hits over 17,704 tokens. Nevertheless, when considering the relative frequency normalized per 10k provided by the #Lancsbox 'Whelk' function,[26] it is Francis Beaumont's *The Masque of the Inner Temple and Gray's Inn* (1613) that has the highest relative frequency (73.48), followed by Jonson's last Stuart masque *Chloridia* (1631) with 56.34, Thomas Campion's *Lord Hay's Masque* (1606) with 55.68, and other masques. The first non-masque on the list of the plays ordered by relative frequency of the lexeme DANCE is *The Ball*, in 9th position (17.51) after eight masques and drolls by Jonson and other contemporary playwrights concerned with masques. Therefore, an important feature of the dance-related lexicon emerges from the quantitative data quoted above: as expected, the masque is the dramatic genre most involved in dance discourse. This high frequency is inversely proportional to the kind of dances actually mentioned in masques, as Ravelhofer also pointed out (2006). Indeed, as seen in other sections of this part, precise mentions of dance names are quite rare in masques, probably because the type of dance performed also depended on the terpsichorean skills of the audience participating in the theatrical event.

Lastly, it is worth analysing a few statistics related to the collocational patterning of the lexeme DANCE. This has been possible, thanks to GraphColl, by setting a span of 1 right and 1 left collocate (and colligate) – since I was interested in very narrow collocations – and a threshold of a minimum of 5 occurrences. Thus the 1,767 occurrences of the lexeme DANCE collocate with 116 content or function words. The most frequent co-occurrence is of the conjunction 'and' (285 hits on the right), which probably means that dancing is often associated with other activities such as music, singing, etc. Indeed, 'sing' and 'singing' are the most frequent content/lexical words co-occurring with 'danc*' within in the EMEP corpus. 'Sing' is always on the right, 'singing' on the left, thus also indicating the tendency of early modern English playwrights to adopt collocations such as 'dance and sing' or 'dance, sing', while preferring 'singing and dancing' or 'singing, dancing', probably for phonological and prosodical reasons. However, this datum is extremely important in understanding the close connection between dancing and singing in early modern English drama, as Eubanks Winkler noted (2020, 1–38). On the other hand, in terms of multiword units that follow the pattern 'adjectivized noun + dance', the most common are 'Country Dance' and 'Morris Dance'.

G

Galliard

Galliard A light-hearted vigorous court dance performed in triple time, mainly after a pavane. It is thought to have originated in Lombardy. It was particularly popular at the court of England's Elizabeth I, where a variant known as the volta was much loved by the Queen.

Concise Oxford Dictionary of Music

SIR TOBY BELCH: What is thy excellence in a **galliard**, knight?
SIR ANDREW: Faith, I can cut a caper.
SIR TOBY BELCH: And I can cut the mutton to't.
SIR ANDREW: And I think I have the back-trick simply as strong as any man in Illyria.
SIR TOBY BELCH: Wherefore are these things hid? wherefore have these gifts a curtain before'em? are they like to take dust, like Mistress Mall's picture? why dost thou not go to church in a **galliard** and come home in a coranto? My very walk should be a jig; I would not so much as make water but in a sink-a-pace. What dost thou mean? Is it a world to hide virtues in? I did think, by the excellent constitution of thy leg, it was formed under the star of a **galliard**.

(W. Shakespeare, *Twelfth Night*, 1601, 1.3)

RAYBRIGHT: What's he that looks so smickly?
HEALTH: A flounder in a frying-pan, still skipping, one that loves mutton so well, he always carries capers about him; his brains lie in his legs, and his legs serve him to no other use than to do tricks, as if he had bought'em of a juggler, he's an Italian dancer, his name –
POMONA: Signior Lavolta (Messer mio) me tesha all de bella Corantoes, **galliardaes**, piamettaes, capeorettaes, amorettaes dolche dolche to declamante do bona robaes de Tuscana.

(T. Dekker and J. Ford, *The Sun's Darling*, 1624, 2)

MASTER MANLEY: But had you seen my Lord lofty dance the **Galliard**, with such capers, such half capers, such turns o'th toe, and above ground, you would have sworn'em sound for nine Generations, and for Lavoltas, La, la, la, &c. and then the Lady.

LADY BEAUFIELD: How did he use her?

MASTER MANLEY: With wonderful skill, he put his right arm about her, and took her left hand in his, and then he did so touze her with his right thigh and leg, and lift her up so high, and so fast, and so round–

(W. Cavendish and J. Shirley, *The Variety*, 1641–49, 3.1)

References to the Galliard, one of the best-known dances in early modern England – probably because Queen Elizabeth was fond of it – amount to 50 (8 of which indicate the proper noun of the French dancing master Galliard in William Cavendish and Shirley's *The Variety*). This highlights the fact that it was a widespread choreography on the early modern stage, and consequently early modern playwrights demonstrate a clear understanding of its execution and main steps. Overlooking the Galliard's fast rhythm and movements, which associate it with other sixteenth- and seventeenth-century dances, the lines quoted above from Shakespeare, Dekker and Ford, and William Cavendish and Shirley illustrate the more interesting and peculiar aspects of such a dance.

Shakespeare's mention of a Galliard in *Twelfth Night* is certainly one of the best known in the entire corpus of early modern English plays. As argued elsewhere (see Ciambella 2017a; 2020), gender-related issues and matters of power balance are intimately linked with the connotation of the Galliard and of dance in general in this scene. Indeed, the penniless noble and fat Sir Toby Belch tries to convince his tall and slim friend, the socially less powerful Sir Andrew Aguecheek, to go on courting his niece Olivia, instead of giving up and returning to his country; otherwise Sir Toby could not exploit the other's money to go to parties and get drunk. Sir Toby's successful flattery uses Sir Andrew's physical and terpsichorean skills as a lever essentially to obtain financial aid. Nevertheless, this scene is also important from a terpsichorean viewpoint because some steps, jumps, and features of the Galliard are present, thus demonstrating the Bard's knowledge of this dance, probably one of the few he described in his canon. First of all, the audience/reader is informed that capers are part of the choreography of the Galliard. Since the Galliard and the Cinquepace were considered the same dance, the caper – a jump similar to the entrechat in ballet – was probably one of the five basic steps of the choreography. However, Sir Andrew does not understand Sir Toby's culinary pun about the caper – the berry he usually combines with mutton meat – and goes on boasting about his ability to perform the best back-trick ever. The back-trick, also known by the French term *ruade*, is a high jump performed by beating the calves in the air. It can be right or left depending on which leg first lifts backwards from the ground. Lastly, in the final part of the scene quoted above, we read that the Galliard requires a great deal of leg power. Indeed, as Sir Toby affirms, Sir Andrew's leg seems so strong and well-defined that he must have been born under the star of the Galliard.

Particular attention to the steps of the Galliard is also given by William Cavendish and Shirley in *The Variety*, where even a French dancing master named Galliard is present. However, a Galliard is described by Master Manner to Lady Beautiful – the couple is destined to end up together as the play closes – together with La Volta, whose presence and connotations in the EMEP corpus will be considered in the section devoted to it. Manner's Galliard has capers, as in Shakespeare's *Twelfth Night*, half capers and turns of the toe (probably indicating some kind of twirl or pirouette), which indicate the great agility and strength of any performer, both male and female. Such precision in the description of the Galliard's basic steps and possible variations may well denote that it was one of the most executed dances on stage by early modern English actors.

Some references to the Galliard in the EMEP corpus collocate with adjectives or nouns which connect this dance to France (not only William Cavendish and Shirley's *The Variety*, where Monsieur Galliard is a French dancing master, but also in Brome's *The City Wit*, 1629–32, 3.2, and Margaret Cavendish's Restoration comedy *The Wits' Cabal, Part 1*, 1662, 5.5). Nevertheless, the Galliard seems to have Italian origins – as Caroso's and Negri's treatises demonstrate. Its Italian origins are quite evident in Dekker and Ford's masque *The Sun's Darling* (1656), where in extremely broken English the Italian dancer Pomona lists a Galliard among the many dances he can teach.

H

Hay

Hay/hey/haye A country dance dating back to the 1520s, or earlier, and featuring a serpentine formation in which two lines of dancers thread through each other. This movement is featured in other dances, including the farandole.

Oxford Dictionary of Dance

GAVESTON These are not men for me:
I must have wanton poets, pleasant wits,
Musicians, that with touching of a string
May draw the pliant king which way I please.
Music and poetry is his delight;
Therefore I'll have Italian masks by night,
Sweet speeches, comedies, and pleasing shows;
And in the day, when he shall walk abroad,
Like sylvan nymphs my pages shall be clad;
My men, like satyrs grazing on the lawns,
Shall with their goat-feet dance an antic **hay**.

(C. Marlowe, *Edward II*, 1592–93, 1.1)

JULSE I tell you my Lord, coming abruptly as your honour or any else may do to the Prince's chamber, about some ordinary service, a found him in his study, and a company of bottle-nos'd devils dancing the Irish **hay** about him, which on the sudden so startled the poor boy, as a clean lost his wits, and ever since talks thus idle, as your Excellence hath heard him.

(J. Day, *The Law Tricks*, 1608, 4)

PORTUNUS Here all the day, they feast, they sport, and spring;
Now dance the Graces' **Hay**, now Venus Ring:
To which the old musicians play, and sing.

(B. Jonson, *The Fortunate Isles and Their Union*, 1624–25)

According to Hoskins (2005, 23–26), Hay is synonym to Branle/Brawl. However, its occurrences in the EMEP corpus suggest different connotations. First of all, as for other dances analysed thus far, even in this case the KWIC and GraphColl functions helped with the polysemy of the lemma *hay*, which may indicate "[g]rass that has been mown and dried for use as fodder" (*OED*, n.), a wine, an exclamation, or a dance. In fact, when the word indicates the sixteenth-/seventeenth-century folk dance, its occurrences are always preceded by a pre-modifier (either possessives or qualifiers) and the verb 'to dance' in the multiword 'to dance + pre-mod + hay'. Therefore, it has been quite easy to identify those occurrences where the Hay was considered a dance.

In Marlowe's *Edward II*, the Hay is defined as 'antic'[27] – i.e., absurd, bizarre, and grotesque – by the king's favourite, Piers Gaveston. Edward I has just died and Gaveston can finally return from exile, so in the initial scene of the tragedy he plans feasts to celebrate to new king who favours him. The Hay is inserted here in a pagan, almost mystical context with nymphs and goat-footed satyrs – thus probably also recalling the atmosphere of Shakespeare's *A Midsummer Night's Dream*. The scene described by Gaveston is grotesque, with satyrs dancing a bizarre Hay with their goat-feet. Edward II is immediately depicted as a weak and manipulable king who loves poetry, music, and dance, this aspect perhaps recalling the Puritan attacks against the art of singing and dancing in the late sixteenth century. Conversely, Gaveston, whom Edward loves so dearly that critics have considered their relationship homosexual (see, for instance, do Desterro 1998; Crewe 2009), plays the role of the morality character Vice who "seduces the soul of the eponymous character" (Kelly 1998, 11).

The Hay is again associated with non-Christian environments in John Day's *The Law Tricks* where, exactly as in Dekker's allegorical pamphlet *A Strange Horse-Race* (1613), Julse tells Ferneze, Duke of Genoa, that he found the latter's son Polymetes in the company of some bottle-nosed devils who dance an Irish Hay. Although, as Winerock suggests (2011, 263), there are no extant manuscripts describing the execution of the Irish Hay, the fact that it is grotesquely performed by some devils both in Day's and Dekker's plays connotates it negatively as a disharmonic dance, far from representing that *musica universalis* which courtly dances signified in early modern English courts.[28]

In the end, a less negative yet always exclusively mystical/pagan connotation of the Hay is present in the last of Jonson's Jacobean masques: *The Fortunate Isles and Their Union*. The Graces' Hay reported by the minor sea god Portunus is associated with Venus's Ring, both dances indicating harmony and circularity, while the Graces and Venus represent beauty. Therefore, both negative and positive connotations of the Hay connect it to a mystical and supernatural world which highlights its ancient pagan origins, "long before the Elizabethan period" (Sorell 1957, 379).

Horn(i/e)pipe

Hornpipe English step dance originally accompanied by a wooden hornpipe (now obsolete). In the mid-18th century it became widely associated with sailors and around the same time it changed from triple to duple time.

Oxford Dictionary of Dance

GODS Then round in a circle our sportance must be,
Hold hands in a **hornpipe**, all gallant in glee. *Dance.*
(G. Peele, *The Arraignment of Paris*, 1581–84, 1.3)

Enter three Antics, who dance round, and take SLIPPER *with them.*
SLIPPER I will, my friend[s], and I thank you heartily: pray, keep your courtesy: I am yours in the way of a **hornpipe**. – [*Aside.*] They are strangers, I see, they understand not my language: wee, wee. – Nay, but, my friends, one **hornpipe** further, a refluence back, and two doubles forward: what, not one cross-point against Sundays? What, ho, sirrah, you gome, you with the nose like an eagle, and you be a right Greek, one turn more.
(R. Greene, *James IV*, 1590–91, 4.1)

CAPONI E're I'll lose my dance,
I'll speak to the purpose. I am Sir no Prologue,
But in plain terms must tell you, we are provided
Of a lusty **Hornepipe**.
COSIMO Prethee let us have it,
For we grow dull.
(P. Massinger, *The Great Duke of Florence*, 1627, 4.2)

KING Sweet, be content.
ALINDA Content yourself, great sir,
With your black infamy; sit down content
On your majestic throne, the president
Of capital contented cuckolds, do,
Till all your subjects dance the **hornpipe** too.
KING Nay, dear Alinda, do but think–
ALINDA Think what?
What? On a course to be revenged on you?
To serve you in that kind myself?
KING Oh, torment!
(R. Brome, *The Queen and Concubine*, 1635–40, 3.3)

The Hornpipe is an ancient folk dance of British origin, and this is also proved by the fact that it appears to be already lexicalised in the EMEP corpus: indeed, unlike other dances,[29] its occurrences are not always necessarily introduced by the transitive verb 'to dance'.

The semantic neighbourhood of the occurrence of the Hornpipe in the gods and muses' performance in Peele's *The Arraignment of Paris* offers important information about the directions of this dance. First of all, the popular character of the choreography is highlighted by the fact that it is danced by "Country Gods", as the initial stage direction reads in 1.3, thus "minor Gods and Goddesses", as stated in the dramatis personae, distinguish this group of deities from the "Olympian Gods and Goddesses". It is not surprising that, being a folk dance, the Hornpipe is typically a round, circular, non-hierarchized dance – hence the collocation "round in a circle" – perhaps performed around a symbolic object, as will be seen when dealing with such dances as the Maypole. As with other circular dances in the EMEP corpus, circular dances are performed by taking each other's hand; thus the collocation "hold hands" or similar ones are found quite commonly as right/left collocate when a circular choreography is about to be danced. Think, for instance, of "[t]he weird sisters hand in hand" in Shakespeare's *Macbeth* 1.3, or "let's take all hands" and "[a]ll take hands", as Antony and Enobarbus repeatedly invite all the onstage characters to perform the Egyptian Bacchanal in *Antony and Cleopatra* 2.7 – two tragedies which, interestingly enough, were probably written in the same year, 1606.

Slipper's Hornpipes at the end of 2.2 and 4.1 in Robert Greene's *The Scottish History of James IV* are probably the most important occurrences of this dance in the EMEP corpus. Indeed, not only does the tragedy associate the Hornpipe to Scotland and popular environments (Slipper is the disgraced courtier Bohan's son), but it also contains some information about the execution of this dance: one step back and two forward.[30] In *James IV*, the energetic Hornpipe is associated with a chaotic moment when the naïve Slipper is robbed by Andrew of all his money, so that his dreams of becoming a courtier like his father vanish when he cannot finish to pay the tailor, the shoemaker, and the cutler.

Vitality and energy are also associated with the lusty Hornpipe[31] offered by the servant Caponi[32] to Cosimo de' Medici in Massinger's early Caroline tragicomedy *The Great Duke of Florence*. Just as Massinger's plot is completely ahistorical, citing a Hornpipe in a play set in Renaissance Italy is out of place. And yet here again the Hornpipe is associated with chaos and deception, since the dance is only a distraction to prevent Cosimo from meeting Lidia, his nephew Giovanni's beloved, to whom the Duke feels a certain attraction.

Deception is often connected with the Hornpipe insofar as this dance is also linked to cuckoldry – as its name indicates. In fact, the ancient Hornpipe is thought to have been performed with a horn, a sort of cornucopia, as a kind of propitiatory choreography before and during harvest periods. Thus the horn from which the dance takes its name then came to be associated with cuckoldry. It is exactly this negative connotation (see also Squirril's "Lancashire hornpipe" with his master Maybery's wife, a nod to the sexual intercourse in Dekker and Webster's *Northward Ho*, 1605–6, 1.3) that characterises the Hornpipe mentioned by the ambitious brand-new queen Alinda in Brome's tragicomedy *The Queen and Concubine*.[33] Here Alinda attacks the Sicilian King Gonzago's honour as a man and a husband by also establishing a metaphorical and allegorical connection between the microcosm of the Court and the macrocosm of all the Sicilian subjects who, following the example of a cuckolded king, will also dance the Hornpipe. Queen Alinda is not talking about herself, of course, but about the innocent former Queen Eulalia whom she is trying to have killed by royal decree and who, according to the vile Queen, committed "foul adultery" against Gonzago.

To conclude, the Hornpipe's association with deception and cuckoldry in the EMEP corpus runs in tandem with the vigour and energy that this dance requires, thus possibly mirroring the same vigour and energy required in the sexual act (this time considered from an extramarital perspective).

J

Jig/Gig/Gigue

An old British folk dance. It may have derived from either the French gigue or the Italian giga. A fast solo dance, it is usually performed in 6/8 or 12/8 time and is characterized by its lively footwork.

Oxford Dictionary of Dance

OBERON: Oberon, King of fairies, that loves thee because thou hatest the world; and to gratulate thee, I brought these antics to show thee some sport in dancing, which thou hast loved well.

BOHAN: Ha, ha, ha! Thinkest thou those puppets can please me? whay, I have two sons, that with one Scottish **jig** shall break the neck of thy antics.

OBERON: That would I fain see.

BOHAN: Why, thou shalt. – Ho, boys!
Enter SLIPPER and NANO.
Haud your clacks, lads, trattle not for thy life, but gather up your legs, and dance me forthwith a **jig** worth the sight.

SLIPPER: Why, I must talk, an I die for't: wherefore was my tongue made?

BOHAN: Prattle, an thou darest, ene word more, and ais dab this whinyard in thy wemb.

OBERON: Be quiet, Bohan. I'll strike him dumb, and his brother too: their talk shall not hinder our **jig**. – Fall to it; dance, I say, man.

BOHAN: Dance Humer, dance, ay rid thee.
The two dance a ***jig*** *devised for the nonst.*

(R. Greene, *James IV*, 1590–91, 1.1)

CARLO BUFFONE: O, do you know me, man? why, therein lies the syrup of the jest; it's a project, a designment of his own, a thing studied, and rehearst as ordinarily at his coming from hawking or hunting, as a **jig** after a play.

SOGLIARDO: Ay, e'en like your **jig**, sir.

(B. Jonson, *Every Man Out of His Humour*, 1599–1600, 2.1)

The Jig, exactly as the Galliard, was one of the most widespread dances in early modern England; hence, the occurrences of the strings 'jig*' and 'gig*' are numerous and the most interesting ones must be carefully selected. First of all, it is worth noting that two of the occurrences of the lemma *Jig* collocate with the adjective 'merry' (one in Middleton's *No Wit, No Help Like a Woman's*, 1653, 1.1, and the other in Brome's *The Court Beggar*, Epilogue), this noun phrase underlining the lively character of this dance.

Moreover, the two scenes quoted above focus our attention on other two important peculiarities of this dance. In Greene's *James IV*, for instance, Bohan praises his two sons Slipper and Nano for their astonishing ability to dance the Jig better than Oberon's Antics. In this scene, the Jig is defined as Scottish – the play is set in Scotland – as in Shakespeare's *Much Ado*, 2.1, and in Margaret Cavendish's *The Wits' Cabal*, 5.5, this geographical indication possibly ascertaining that the Jig danced in the British Isles probably comes from Scotland, which John Playford also seems to suggest in his *Apollo's Banquet* (1663). Nevertheless, Irish Jigs were also famous at that time in England, although the EMEP corpus appears to contain no reference to them.

Another important aspect which emerges from the corpus analysed is that Jigs were also performed at the end of early modern English plays.[34] Such information, as noted in Part I, can be found in other sources that are not theatrical – e.g., Thomas Platter's diary or the *Order for Suppressing Jigs at the End of Plays* (1612) – but even plays seem to suggest that such activities were quite common.[35] In Jonson's *Every Man Out of His Humour*, the jester Carlo Buffone affirms that it is absolutely normal to have a Jig after the end of plays and, implicitly, that the jester performs it (think, for instance, of stage Jigs performed by famous early modern clowns such as Tarlton or Kempe. Cf. Baskervill 1929, 87). The same considerations are present in Marston's *Jack Drum's Entertainment* (1601, 1), where the foolish courtier John Ellis says that "the Jig is called for when the play is done", once more associating the Jig with clownish/foolish characters who might have performed it at the end of sixteenth- and seventeenth-century plays.

L
(La)Volta

Volta [It. and Fr., turn] A fast dance for couples performed in simple triple time, it is related to the Italian galliard. It features quick, leaping turns during which the woman is lifted by her male partner. A lusty dance, considered by some to be immoral, it was a favourite of Queen Elizabeth I. In Britten's opera Gloriana the volta is dance by Elizabeth I and her court.

Oxford Dictionary of Dance

PHANTASTES: List, how my heart envies my happy ears! Hisht, by the gold strung harp of Apollo, I hear the celestial music of the spheres, as plainly as ever Pythagoras did; O most excellent diapason, good, good, good! It plays Fortune, my foe, as distinctly as may be.

COMMUNIS SENSUS: As the fool thinketh, so the Bell clinketh; I protest I hear no more than a post.

PHANTASTES: What, the **Lavolta** hay? Nay, if the heavens fiddle, Phansy must needs dance.

COMMUNIS SENSUS: Prethee sit still, thou must dance nothing but the passing measures. Memory, do you hear this harmony of the spheres?

MEMORIA: Not now my Lord, but I remember about some 4000 years ago, when the sky was first made, we heard very perfectly.

ANAMNESTES: By the same token the first tune the planets played, I remember Venus the treble ran sweet division upon Saturn the base: The first tune they played was Sellenger's round, in memory whereof ever since it hath been called the beginning of the world.

(T. Tomkis, *Lingua*, 1607, 3.7)

THISBE I will tell thee,
You know Feledemus the dancer
ARSINOE Yes.
THISBE We too must dance **la volta** here tonight,
Let's have a chamber and a bed made ready
Sweet sister, for I've promis'd him this night
He shall come first, and I will follow after,

As soon as I to bed have brought my mistress.
ARSINOE All things shall be according to thy wish,
My better half.
THISBE But when he comes I pray you
Be out the way, for he is very shame-fac'd
As being a novice in this art.
ARSINOE Is he
Not entered yet?
THISBE Not yet I can assure yee,
His flower is yet to spend and strength of youth.
(J. Gough, *The Strange Discovery*, 1640, 3.5)

MASTER MANLEY: But had you seen my Lord lofty dance the Galliard, with such capers, such half capers, such turns o'th toe, and above ground, you would have sworn'em sound for nine Generations, and for **Lavoltas**, La, la, la, etc. and then the Lady.
LADY BEAUFIELD: How did he use her?
MASTER MANLEY: With wonderful skill, he put his right arm about her, and took her left hand in his, and then he did so touze her with his right thigh and leg, and lift her up so high, and so fast, and so round–
(W. Cavendish and J. Shirley, *The Variety*, 1641–49, 3.1)

The occurrences of La Volta as a dance in the EMEP corpus were extremely difficult to detect, due to the highly fluctuating spelling of the lexeme and word forms sought. Nevertheless, the string 'la**lt*' provided the most pertinent and statistically significant results. The GraphColl tool highlighted some interesting close collocates, such as 'high' (e.g., Shakespeare's *Henry V*, 3.6, *Troilus and Cressida*, 1602, 4.4, Heywood's *The Rape of Lucrece*, 1608), or 'leap' (e.g., Marston's *The Fawn*, 1604, 2), which immediately underline the most obvious characteristic of this dance: the fact that women were lifted in the air while the dancing couples were turning around the stage or ballroom. A very detailed stage direction for the execution of La Volta is given by William Cavendish and Shirley in *The Variety*. In the scene previously considered in discussing the Galliard, Master Manley lists to Lady Beautiful the main steps of La Volta, and he does so by using such body language that the physicality of this dance is immediately perceivable. The execution of La Volta as described in *The Variety* requires great coordination and close proximity of the two partners' bodies – which is one of the reasons the Puritans railed against what they saw as such a lewd performance. This negative connotation of La Volta is also present in Richard Nathaniel's *Messalina* (1639–40, 3.2), when Montano states that this dance mirrors the "whorish fortitude" of its performers, as also noted by Hopkins (2016, 136).

A strong sexual connotation of La Volta is present in John Gough's *The Strange Discovery*, where dancing La Volta is an indisputable metaphor for the sexual act. The scene is full of clear double entendres aimed at revealing the virgin Thisbe's sexual intercourse with Feledemus who, not coincidentally, is a dancer that has "not entered yet" and so will dance a La Volta with her in bed for the first time. As in many of the occurrences of lively dances analysed thus far, even La Volta has close connections with sex, since it requires intimate physical proximity, strength, and coordination.

Another significant feature of this dance is its circularity, both within the dancing couple (since women are lifted in the air while turning; hence the name La Volta, from the Italian 'voltare' = to turn) and in their circular disposition on the stage/ballroom. It is exactly this circularity which is recalled in Thomas Tomkis's academic allegorical comedy *Lingua, or the Combat of the Tongue and the Five Senses for Superiority*. La Volta and the Passamezzo (or Passing Measure) are the two dances which reproduce the harmony of the celestial spheres on earth, according to Phantastes and Common Sensus, this connotation contrasting the Puritan and anti-terpsichorean view of a choreography which arouses the males' sexual gaze. Tomkis's view is much more oriented to Neoplatonic considerations which see the microcosm of dance as the perfect embodiment of the macrocosm of the harmony of the spheres.

M

Maypole Dance

> Maypole dance, ceremonial folk dance performed around a tall pole garlanded with greenery or flowers and often hung with ribbons that are woven into complex patterns by the dancers. Such dances are survivals of ancient dances around a living tree as part of spring rites to ensure fertility. Typically performed on May Day (May 1), they also occur at midsummer in Scandinavia and at other festivals elsewhere. They are widely distributed through Europe – e.g., "Sellenger's Round" in England, the *baile del cordón* of Spain – and also are found in India. Similar ribbon dances were performed in pre-Columbian Latin America and were later integrated into ritual dances of Hispanic origin. Maypoles may also appear in other ritual dances, as in the Basque *ezpata dantza*, or sword dance.
>
> *Encyclopædia Britannica*

No stage directions for the Maypole Dance are introduced in early modern plays. This is probably due either to the fact that there were no precise instructions about the steps to perform, or that no particular choreographic directions needed to be followed. Indeed, rather than a dance per se, the Maypole is a kind of folk propitiatory rite performed on May Day around the May pole, a tall trunk decorated with flowers, branches, leaves, and colourful ribbons which were used choreographically by the dancers. In other words, people performed the Maypole Dance simply by turning around the May pole and weaving the ribbons together.

What is interesting to note in this section are the collocations of the lemma *maypole* or the chunk 'May Pole' within the EMEP corpus. Many occurrences collocate with 1L proper nouns such as Banbury, Strand, etc., thus indicating specific May poles placed in various cities, villages, or quarters. As for the collocational patterning that #Lancsbox provided, there is only one occurrence of the multiword 'May Pole' which refers unequivocally to the dance and it is in Thomas Nabbes's comedy *Covent Garden* (1632–38, 1.1): "I had rather see a Morris dance and a May pole, then ten plays", says Dobson, Dungworth's servant. Similar to what has been stated about the Hay's collocational patterning, even the lexeme MAYPOLE has its own specific co-text formed by the multiword 'verb (mainly dance or revel) + about + MAYPOLE', which does not indicate a precise choreography, as stated above, but the action of performing it around the

May pole (e.g., Shirley's comedy *The Constant Maid*, 1640, 2: "I've a great mind to dance about a Maypole; shall we?"; and Robert Cox's droll *Acteon and Diana*, 1656, 1: "This day thou knowest the maids and young men meet to sport, and revel it about the May-pole").

Measure

> **Measure.** A stately Eng. dance of 15th and 16th cents. ('trod a measure' is a frequent phrase in Elizabethan drama).
>
> *Concise Oxford Dictionary of Music*

LUDIO The highest fixed stars, what do they all the year but dance their **measures**, and keep therein their stately distances, and, to shew their mirth and wantonness, they with their glaring eyes twinkle on their spectators. The Planets have varieties of dances, sometime grave Pavans, otherwhile nimble Galliards.

(W. Hawkins, *Apollo Shroving*, 1626, 5.4)

EULINUS Though Orpheus' harp, Arion's lute, the chimes
Whose silver sound did Theban towers raise,
Though sweet Urania with her ten-string lyre,
Unto whose stroke the daily-rolling spheres
Dance their just **measures**, should with tune and tone
Tickle my air-bred ear, yet can their notes
Those fabulous stones more enter than my soul.

(J. Fisher, *Fuimus Troes*, 1633, 4.2)

Song.
[…]

Rose-colour'd Modesty and Truth
Dance harmless **measures** in this place,
With health and a perpetual youth,
And all your virgin trophies bring away
To grace these nuptials! Triumph's holyday!

A dance.

TRIVULCI You have our hearty thanks, and we shall study
To give your fair requital. Come, my lord!
Erect your drowsy spirits; let your soul
Dance airy **measures** in your jocund breast.

(H. Glapthorne, *The Lady's Privilege*, 1640, 5.1)

The lemma *measure* is an extremely difficult one to deal with in early modern plays. Aside from its polysemy and various connotations, when talking exclusively about dance, at least three different meanings can be found. First of all, the umbrella term 'Measure' can be used as a synonym for dance (see, for instance, the list of the eight Old Measures performed during the revels at the Inns of Court, where the term 'measure' stands for dance, choreography). Such primary meaning is in a hypernym of Measure meant as Pavan (Hoskins 2005, 63–78), which in turn is a hypernym of Measure understood as a particular movement or pattern of choreography. This complex and multi-layered hypernymic/hyponym relation is reflected in the many occurrences of the lexeme MEASURE in the EMEP corpus (770 hits in 344/666 texts). Nevertheless, the GraphColl function has highlighted some recurring collocations within the semantic field of astronomy, the most interesting of which have been reported above. None of the other dances inserted or dealt with in early modern English plays seem to have such a close connection with the movement of the celestial spheres as the Measure. This means that, whatever the hypernymic or hyponymic connotation of the lexeme, the Measure stands for a (generic) courtly dance whose ordered and hierarchic positions and movements reproduce those of the celestial macrocosm on the earthly microcosm of the Elizabethan, Jacobean, and Caroline courts.

In Hawkins's *Apollo Shroving*, 'measure' is certainly used as an umbrella term to indicate any kind of courtly dance which metaphorically and metonymically reproduces the revolution of the celestial orbs. Indeed, the fixed stars are said to dance their Measures yearly and the planets dance grave Pavans or nimble Galliards. In addition to providing important information about astronomical conceptions in seventeenth-century England (see Ciambella 2017b), this scene clearly presents the Measures as slow-paced stately dances[36] with unchanging hierarchised positions.

The same connection between the Measure and the celestial sphere is introduced by the noble Eulinus in Jasper Fisher's verse drama *Fuimus Troes. The True Trojans*. As Butler argues (2007, online), Eulinus's utterances are imbued with "Neoplatonic excess", and the assertion that "the daily-rolling spheres / Dance their just measures" perfectly fits the Neoplatonic principle of the *musica universalis*. Eulinus's Measures are also 'just', an adjective which in early modern English indicated that someone or something was "morally upright, righteous in the eyes of God" or "fitting, proper, conforming to standards or rules" (*Oxford Dictionary of English Etymology*), a connotation that adds religious nuances and legitimacy to the dancing of Measures during a reign where the dance-addicted Queen Consort was also a Catholic.[37]

Both the song lyrics and the Duke of Trivulci's speech in Henry Glapthorne's *The Lady's Privilege* offer interesting sparks for a collocational analysis of the semantic neighbourhood of the lexeme MEASURE. First of all, these Measures are defined as 'harmless', whose meaning of "without power or disposition to harm" has been attested since the 1530s, according to the *Oxford Dictionary of English Etymology*. The scene is full of qualifiers belonging to the semantic sphere of grace and harmony; hence, even the Measures quoted by the Duke are 'airy', an adjective that since the 1620s had acquired the connotation of both "done in the air" and "sprightly, light in movement" (*Oxford Dictionary of English Etymology*). Therefore, despite not being strictly connected to the harmony of the celestial spheres, in Glapthorne's *The Lady's Privilege*, the Measure is presented as a graceful courtly dance particularly suited to celebrating weddings.

Moresca/Morisco

> **Moresca** (It., Sp.), Moresque (Fr.) A dance popular throughout Europe in the 15th and 16th centuries. Its performers wore Moorish costumes, had blackened faces and bells on their legs. The most popular dance of Renaissance ballets and mummeries, it frequently involved a mock sword fight between 'Christians' and 'Moors', also known as a *danse des bouffons*. The *moresca* survives in Spain, Corsica, and Guatemala and is probably related to the English morris dance.
>
> *Concise Oxford Dictionary of Music*

> CLEM I am so tir'd with dancing with these same black she-chimney sweepers that I can scarce set the best leg forward; they have so tir'd me with their **moriscos**, and I have so tickled them with our country dances, Sellenger's round and Tom Tiler. We have so fiddled it!
>
> (T. Heywood, *The Fair Maid of the West, Part 2*, 1631, 2.1)

> MASTER MANLEY And now he must be in Cuerpo, or like a fellow on the ropes, or a Tumbler when he shoots his body through a hoop; there was music then, and a Heaven and Earth, beyond your brawls, or your Mountague, with a la, la, la, like a Bacchanalian dancing the Spanish **Morisco**, with knackers at his fingers.
>
> (W. Cavendish and J. Shirley, *The Variety*, 1641–49, 3.1)

Few dances had a worse connotation than the Moresca (or Morisco), which raises essential racial and religious issues in early modern England. In fact, the Moresca is named after the Moors, Arabs who had conquered Spain in the 8th century CE. However, in sixteenth- and seventeenth-century England, since the Spanish Moors were the nearest Muslims to Western Europe, any Muslim of a dark complexion was termed a Moor (Shakespeare's Othello, the Moor of Venice, is probably the most famous example of this semantic twist/amplification). Taking into account all the xenophobic considerations expressed in examining the English Country Dance and its difference from the Moresca in Heywood's *The Fair Maid of the West*, I would add here that the Moresca was perceived as a dance belonging to the Islamic invaders who had conquered part of Europe and brought their religion into the Continent. For this reason, occurrences of the Moresca in the EMEP corpus are preceded by such qualifiers as 'mad' or 'wild' (as in Shakespeare's *Henry VI, Part 2*, 1590, 3.1), which, from a terpsichorean point of view, also highlight the boisterous and chaotic character of this dance, one that is not associated with courtly environments.

In William Cavendish and Shirley's *The Variety*, the Moresca is associated both with its geographical origin and with the kind of dancer who perfectly suits it: a drunkard. Indeed, in the 1600s, the adjective Bacchanalian – substantivized in this case – denoted a person/situation "characterized by intemperate drinking" (*Oxford Dictionary of English Etymology*), someone whom Master Manley considers the perfect performer for a Spanish Moresca. Nevertheless, executing the Moresca must have been anything but simple. In fact, it was often performed as a warlike dance between Moors and Europeans, as if it were a veritable dumb show.

Morris Dance

> **Morris Dance** English ceremonial folk dance which first appeared in England in the 15th century. Its origins are unknown, although it may have derived from the moresca, a dance found in Burgundy in the early 1400s. Traditionally performed by men wearing bells tied to their legs, it is composed of intricate steps and is usually danced in 2/4 time, although it can also be danced in 3/4 time. The dancers may be made up to represent particular characters, such as Fool or Maid Marian, and a cardboard horse is also a regular feature. Some of the elements of morris dancing were used by Ashton in *La Fille mal gardée*.
>
> *Oxford Dictionary of Dance*

Enter Ralph, dressed as a May-lord.

RALPH London, to thee I do present the merry month of May; [...] For now the fragrant flowers do spring and sprout in seemly sort, The little birds do sit and sing, the lambs do make fine sport; And now the birchen-tree doth bud, that makes the schoolboy cry; The **morris** rings, while hobby-horse doth foot it feateously.

(F. Beaumont, *The Knight of the Burning Pestle*, 1607, 4.5)

JAILER'S DAUGHTER: You never saw him dance?
JAILER: No.
DAUGHTER: I have often. He dances very finely, very gracefully.
JAILER: That's fine, indeed.
DAUGHTER: He'll dance the **Morris** twenty mile an hour!

(J. Fletcher and W. Shakespeare, *The Two Noble Kinsmen*, 1613, 5.4)

MOLL: But birds of a feather will fly together; and you and he are seldom asunder.
ALEXANDER: Why, you young witch, call your elder brother fool! But go thy ways, and keep thy maidenhead till it grow more deservedly despised than are the old base boots of a half-stewed pander: lead a Welsh **morris** with the apes in hell amongst the little devils; or when thou shall lie sighing by the side of some rich fool, remember, thou thing of thread and needles, not worth three-pence halfpenny.

(W. Rowley, *A Match at Midnight*, 1622, 3.1)

The Morris Dance is certainly one of the most mysterious dances in early modern England. On stage, one of its first performances dates back to 1514, during Henry VIII's reign, in William Cornish's interlude *The Triumph of Love and Beauty*. The origins of the Morris Dance are uncertain; it is said to derive from the Moresca (given its warlike execution), but when it reached the British Isles it was enriched with so many allegorical values (the legend of Robin Hood and Maid Marian, the May Day rites, etc.) that it is quite difficult to properly retrace its history and origins (see, among others, Forrest 1999; Cutting 2005). Undoubtedly, as seen in Part I, it was William Kempe's pamphlet *Nine Days Wonder* (1600) that contributed to the vast spreading of this dance in late Elizabethan, then Jacobean and Caroline drama, with allusions and performances which underline its allegorical connotation. In Beaumont's *The Knight of the Burning Pestle*, the Morris Dance is associated with the May Day festival, this feature underlining the intimate connection and folk character of both the Morris and Maypole dances. The Citizen's apprentice Ralph enters the stage dressed up like a May lord, a young man who presides over the May Day celebrations, whose figure and role date back to the Celtic May Day festival of Bentane. Ralph delivers a long monologue about the festival and also quotes the Morris Dance. What is worth noting about this quotation is that the Morris is defined as a "ring", which recalls the symbolic circularity of this ritual dance.

From a linguistic point of view, the strong connection between the Morris Dance and circular performances is so evident, that mentions of this choreography are often preceded by the intransitive structure 'dance in', instead of the transitive verb 'dance', as with the majority of the dances dealt with in this book. Indeed, according to Trapdoor all the city whores "dance in a morris" in Dekker and Middleton's *The Roaring Girl* (1607–10, 1.2), and the Second Clown's fore-gallant is "in a morris" in Dekker, Ford, and Rowley's tragicomedy *The Witch of Edmonton* (1621, 2.1), as well as Mr. Plenty's lady in Massinger's *The City Madam* (1632, 2.2).

As stated above, Kempe contributed significantly to popularising the folk Morris Dance in the early seventeenth century. In his *Nine Days Wonder*, Shakespeare's (former) clown performs a Morris Dance from London to Norwich in nine days simply to prove to Queen Elizabeth that he is still vigorous and strong enough to perform at her Court. This energetic performance and the distance covered by Kempe must have influenced the Jailer's daughter's utterance, "He'll dance the Morris twenty miles an hour", in Fletcher and Shakespeare's *The Two Noble Kinsmen*. In the comic scene analysed here, the Doctor suggests that a Wooer, disguised as Palamon, the Jailer's daughter's beloved, make love to her in order to cure her melancholy. Her father seems to agree and among the many qualities his daughter says Palamon possesses, there is his ability to perform the Morris Dance very rapidly. Of course, this is a pun intended to mask the girl's fantasy (or maybe certainty, as the Doctor says) that the boy is a great lover. As seen elsewhere, even in this case the energy and liveliness of a dance are associated with a character's sexual ability – whether proven or not.

In the end, the only adjective of nationality collocating with 'Morris' that #Lancsbox discerned is in Rowley's *A Match at Midnight*. Among the insults that Alexander Bloodhound addresses to his sister Moll, there is also a Welsh Morris Dance. Welsh Morris Dances were not significantly different from their English counterparts, nor were they a particular kind of tune (at least until 1735, when John Walsh inserted a dance called The Welsh Morris in his *The Third Book of the Complete Country Dancing-Master*). Therefore, Alexander's mention of a Welsh Morris might be due to the fact that one of his sister's suitors is Randall the Welshman or, given the rather high tone of their conversation, the use of Welsh may be associated with the etymology of this adjective, which originally meant foreign, not free or servile. In Alexander's terrific speech the imagery associated to the Morris Dance becomes a devilish hallucination of insults wherein his sister is condemned to dance a chaotic choreography with the apes and devils of Hell, which also somehow recalls the pagan, non-Christian origin of the Morris.

P

Passamezzo

Passamezzo (It.) Dance in quicker tempo. It. dance of 16th and 17th centuries similar to [the] pavan but faster and less serious. Examples by Byrd and Philips are in [the] Fitzwilliam Virginal Book.

Concise Oxford Dictionary of Music

SIR TOBY BELCH: Sot, didst see Dick surgeon, sot?
CLOWN: O, he's drunk, Sir Toby, an hour agone; his eyes were set at eight i' the morning.
SIR TOBY BELCH: Then he's a rogue, and a **passy measures** pavan: I hate a drunken rogue.

(W. Shakespeare, *Twelfth Night*, 1601, 5.1)

COMMUNIS SENSUS: Prethee sit still, thou must dance nothing but the **passing measures**. Memory, do you hear this harmony of the spheres?
MEMORIA: Not now my Lord, but I remember about some 4000 years ago, when the skie was first made, we heard very perfectly.

(T. Tomkis, *Lingua*, 1607, 3.7)

PAGE: Alas'twill kill me, I'm even as full of qualms as heart can bear: how shall I do to hold up? Alas Sir I can dance nothing but ill-favor'dly, a strain or two of **Passa-Measures** Galliard.
SINQUAPACE: Marry y'are forwarder then I conceiv'd you, a toward stripling; enter him Nicholao, for the fool's bashful, as they are all at first till they be once well entered.
USHER: **Passa-Measures** Sir?
SINQUAPACE: I Sir, I hope you hear me; mark him now Boy. *Dance.*

(T. Middleton, *More Dissemblers Besides Women*, c. 1614, 5.1)

The Passamezzo, despite being one of the dances quoted by Caroso and Negri in their terpsichorean treatises, was definitely not one of the most widespread dances in early modern English drama. First of all, as the examples quoted above demonstrate, the Italian term Passamezzo was strongly perceived as foreign; indeed, it was never lexicalised or adopted as a loanword in the corpus analysed here. It is as if the name of this Italian dance entered early modern English by phonological similarity with the verb 'to pass' and the name of another dance, the Measure.

As a consequence, discordant, rather confused connotations of the Passamezzo are present in the EMEP corpus. According to Dobson and Wells (2001, 337), in Shakespeare's *Twelfth Night* the surgeon Dick is insulted as "a passy measure pavan" by Sir Toby simply because the Passamezzo has an eight-bar phrase structure and the doctor has his eyes set "at eight i' the morning". A more convincing explanation was given by E. Cobham Brewer in his *Dictionary of Phrase and Fable* (1894); he defined it as follows: "passy-measure pavin [...] a reeling dance or motion, like that of a drunken man, from side to side" (661), given that the Passamezzo was faster than the Pavan. In other words, in *Twelfth Night*, the Passamezzo is associated with drunkenness, exactly as are such dances as the Hay or the Moresca.

A diametrically opposite connotation is present in Tomkis's *Lingua*, where Communis Sensus lists the Passamezzo as one of the best candidate dances to reproduce the harmony of the celestial spheres. Since Memoria affirms that she heard the *musica universalis* about 4000 years ago, when the sky was created, Communis Sensus suggests that it is time they dance anything but the Passamezzo to hear the celestial music again. Therefore, unless the connotations of the Passamezzo changed unexpectedly from the reign of Elizabeth Tutor to that of James Stuart – which is highly improbable if not absurd – early modern English playwrights were definitely not keen on the Passamezzo.

To make matters worse, in Middleton's *More Dissemblers Besides Women*, the Passamezzo once again acquires an adjectival function, pre-modifying the noun Galliard this time instead of Pavan, as in Shakespeare. Actually, neither Shakespeare nor Middleton are completely wrong: on the one hand, the Passamezzo was similar to the Pavan – albeit a bit faster – while on the other hand, it was often followed by a more vivacious Galliard, according to both *Treccani* and *De Mauro*, dictionaries of the Italian language. Nevertheless, even in Middleton's *More Dissemblers Besides Women*, the Passamezzo acquires a negative connotation, since it is the only dance that Lactantio's mistress, disguised as his page Antonio, can dance, and not even that well, although the Passamezzo is considered easy perform. The task of the dancing-master Sinquapace – definitely a speaking name – is to teach her how to dance more complex and livelier choreographies.

Pavan

Pavan (Fr. *pavane*; It. *pavana*; old forms include *pavin*, *pavyn*, *paven*, etc.) The pavan was a dance of Italian origin, popular in the 16th and 17th centuries, and as the name sometimes appears as *padovana* it is assumed that its original home was Padua. It was in simple duple time, and of stately character. In Italy the pavan gave way to the passamezzo by the mid-16th century, but was given a new lease of life by its treatment by English composers, e.g. Byrd, Dowland, Bull, and Philips. It was usually paired with the galliard and their association was the origin of the suite. Some 19th- and 20th-century composers have written works to which they gave the name Pavan, e.g. Fauré's *Pavane*, Ravel's *Pavane pour une infante défunte*, and the Pavan in Vaughan Williams's *Job*.

Concise Oxford Dictionary of Music

LAZARILLO I shall be mous'd by pussycats, but I had rather die a dog's death; they have nine lives apiece – like a woman – and they will make it up ten lives if they and I fall a-scratching. Bright Helena of this house, would thy Troy were a-fire, for I am a-cold; or else would I had the Greeks' wooden curtal to ride away! Most ambrosian-lipp'd creature, come away quickly, for this night's lodging lies cold at my heart.

The Spanish ***pavin****.*

The Spanish **pavin**! I thought the devil could not understand Spanish, but since thou art my countryman, O thou tawny [Satan], I will dance after thy pipe.

He dances the Spanish ***pavin****.*

(T. Dekker, *Blurt, Master Constable*, 1602, 4.2)

LUDIO The Planets haue varieties of dances, sometime graue **Pauins**, otherwhile nimble Galliards.

(W. Hawkins, *Apollo Shroving*, 1626, 5.4)

If one agrees that the Pavan and the Measure were the same dance, as Hoskins suggests (2005, 63–78), occurrences of this dance in the EMEP corpus are numerous. Nevertheless, as shown in Part I, the Pavan was one of the eight Old Measures performed during revels at the Inns of Court; for this reason, I deal with the Pavan as a separate dance.

As stated in the dictionary entry above, the Pavan is associated with the northern Italian city of Padua and so is believed to have an

Italian origin. Sometimes, the Pavan is said to take its name from the Italian, French, and/or Spanish term for 'peacock', respectively *pavone*, *paon*, and *pavón* (Sachs 1937, 356). However, when analysing the EMEP corpus using #Lancsbox, it emerges that the only adjective of nationality which collocates with the occurrences of the lexeme PAVAN is 'Spanish'. Lazarillo's performance of the Pavan in Dekker's *Master Constable* is the only one in the entire early modern English dramatic output, at least according to the corpus-driven analysis carried out here. In Dekker's works, the term Pavan is premodified by the adjective Spanish three times, as if it were necessary to specify it each time. Nevertheless, the same co-occurrence of the noun phrase 'Spanish Pavan' can be found in Jonson's Jacobean comedy *The Alchemist* (1610–12, 4.4) and in Margaret Cavendish's *Wits' Cabal* (5.5). Neither Italian nor French Pavans are mentioned in the 666 texts of the EMEP corpus. This datum can be variously interpreted: on the one hand, the need to specify the origin of this dance can mean that when an Italian or French Pavan is mentioned/performed it is not necessary to make that clear. On the other hand, as Grove had it (1900, 676), "[t]he Spanish Pavan [...] was a variation of the original dance".

Moreover, little information has been found about the performance and character of this dance. The only reference to its stateliness (and slow pace) derives from the already mentioned assertion by Ludio in Hawkins's *Apollo Shroving* about the metonymical and metaphorical connection between dance and the harmony of the celestial spheres. The schoolboy defines the Pavan as a grave dance, the adjective indicating a solemn, sober, important, and serious performance from the 1580s-90s, according to the *Oxford Dictionary of English Etymology*.

R
Round(el)/Ring(let)

Round Dance

1. A dance in which the performers turn round.
2. (a more common use of the term) A dance in which they move round in a circle, i.e. a ring dance.

Concise Oxford Dictionary of Music

Song

PSYLLUS O for a wench, (I deal in faces,
And in other daintier things,)
Tickled am I with her embraces,
Fine dancing in such fairy **rings**

(J. Lyly, *Campaspe*, 1580–81, 1.2)

If you accept this dance of a fairy in a **circle**

(J. Lyly, *Sappho and Phao*, 1584, Epilogue)

FIRST WITCH Ay, sir, all this is so: but why
Stands Macbeth thus amazedly?
Come, sisters, cheer we up his sprites,
And show the best of our delights:
I'll charm the air to give a sound,
While you perform your antic **round**:
That this great king may kindly say,
Our duties did his welcome pay.

Music. The witches dance and then vanish, with Hecate.

(W. Shakespeare and T. Middleton, *Macbeth*, 1606/16[38], 4.1)

Since the very first formulation of the notion of collocation by Firth in the 1930s (1957), it has been clear that in lexicography – and especially in corpus-informed lexicography – collocations are a "binary relationship between a node or base, generally a noun or verb, and its collocates, adjective or adverb" (Williams 2001, 66). This is even more evident in specialised corpora – or in a specialised lexicon in this case – where it often happens also that words with a broad, inclusive meaning undergo a process of semantic narrowing, thus becoming to connotate more specific/specialised referents. This is exactly what happens in dance ESP, as seen above, with the lexeme MEASURE, with the results that both the general, inclusive meaning and the restricted ones coexist and establish a semantic relation of hypernymy/hyponymy. In this case, however, the noun 'round', which indicated "a spherical body" beginning in the 1300s, underwent a semantic specialisation (due to analogy) and the word began to indicate a hypernym for any kind of circular dance. For this reason, because it is such an umbrella term for any circular dance, the collocational patterning of this lemma was quite complex to identify. Nevertheless, in a specialised lexicon, the most recurring collocations are those formed by a noun which is usually pre-modified by an adjective (Williams 2001, 67) or, especially in English, by another noun with an attributive function. These premises helped me find interesting collocations formed by 'adjective/noun used attributively + node' related to the Round(el)/Ringlet using #Lancsbox KWIC, but above all using the GraphColl function. The results extracted define at least two different connotations of the lexeme ROUND as a dance which basically embodies most of the connotations that dancing had in early modern England.

On the one hand, courtly round dances are emblems of the circular and harmonic movements of the celestial orbs and, when this connotation emerges, performers of such dances are mainly heavenly, unworldly beings, especially fairies (think of the Carol(e)s and Roundels performed by Oberon and Titania's fairies in Shakespeare's *A Midsummer Night's Dream*, as brilliantly analysed in Hiscock 2018). Fairy Ringlets have been performed onstage or mentioned since the very beginning of the Elizabethan – even pre-Shakespearean – drama, as evidenced by the above quotations from John Lyly's plays. Psyllus's stanza in the song he sings with other characters onstage in *Campaspe* deals with fairy rings, just as the Epilogue of *Sappho and Phao* describe fairies who dance in a circle.

On the other hand, folk dances also have circular choreographies – e.g., the Maypole or Morris Dances analysed earlier – whose ritual, propitiatory character is not so distant from the more refined fairy Rounds. The boisterous and lively Saxon Round in Chapman and/or Peele's *Alphonsus* (3.1) is a propitiatory ritual dance that the page Alexander calls forth to celebrate Edward and Hedewick's marriage. Similarly, the Round performed by soldiers in Middleton's *More Dissemblers Besides Women* (3.2) acquires the metaphorical connotation of a good omen.

An evil omen is symbolized by the witches' antic Round in Shakespeare's *Macbeth* – the scene taken into account here was actually attributed to Middleton's 1616 interpolations. As this is actually the only occurrence of the collocation 'antic Round' in the EMEP corpus, it is worth taking it into account. The antic Round scene can be read as an antimasque (see Russel Brown 1982, 56), a mystical, ill-fated show, almost a parodic version of Lyly's fairy circles. It is antic, exactly as Marlowe's "antic Hay" in *Edward II* (1.1), hence it acquires the same connotation of bizarre and grotesque performance, a disharmonic Round by the "weird sisters", three bearded witches "so wither'd and so wild in their attire", as Banquo describes them in 1.3. Therefore, the ominous antic Round is the appropriate conclusion to the destabilising dumb show that Hecate and the witches force Macbeth to watch recounting his defeat and Banquo's descendants' success.

Notes

1 For a thorough study of the Almain and its dissemination in early modern England, see Payne 2003.
2 Dates of (estimated) composition/publication are taken from the VEP Early Modern Drama Collection and/or EEBO-TCP. Given the periodization issues presented in the Introduction, when available the presumed date of composition is preferred. Moreover, all quotations from early modern texts are taken from the corpus created or from EEBO-TCP; hence, no additional bibliographical sources are acknowledged; in order to facilitate the legibility of this Part of the book and to avoid confusion, verse/line numbers are not indicated. Similarly, in terms of spelling modernization, all the plays belonging to EMEP corpus, EEBO-TCP reference texts and their titles have been normalized according to the VEP norms (see https://graphics.cs.wisc.edu/WP/vep/variation/).
3 Issues surrounding authorship are much debated in this case. See Blamires 2018 and Jackson 2019 for an overview of the attribution of this play to Chapman and/or Peele. Since matters of authorship are irrelevant to the purposes of our study – at least in this section – I hold that both playwrights may have contributed to the writing of the tragedy. On the other hand, I reject Wiggin's guess (2012) of 1630 as the tragedy's date of composition, also because the Almain was definitely out of fashion during the Caroline era, as even the few references to it on EEBO-TCP demonstrate.
4 When referring to lexemes, it is common practice to use small caps. In addition, I indicate lemmas in italics and specific parts of speech and emphasised words in single quotes.
5 First names of the various playwrights are given only in the very first occurrence here. Nevertheless, to better distinguish between the husband and wife couple of the Duke and Duchess Cavendish, and their respective plays, their first names (William and Margaret) are always noted.
6 Five different authors belonging to the Inns of Court are responsible for the five acts of the play. They are: Rod. Staf., Hen. No. (probably Henry Noel), G. Al., Ch. Hat. (perhaps Christopher Hatton); and R. W. (Robert Wilmot, who printed the tragedy for the first time in 1591).
7 For instance, Negri's *Le gratie d'amore* lists a choreography called "L'Alemana d'amore" (the Almand of love).
8 According to the *OED*, the lemma *Almain* indicates "a native or inhabitant of Germany or the lands corresponding to modern Germany; a German.

Chiefly archaic and literary in later use" (n. 1) or "of, relating to, or characteristic of Germany or the German language; German" (adj.).

9 See also Shakespeare's *Hamlet* 1.4.8, where Claudius is said to "keep wassail, and the swaggering up-spring reels".

10 Looking for words or multiword units by inserting wildcards helps users of corpus linguistic tools to easily identify possible spelling variations, inasmuch as the wildcard function is thought to replace the asterisk with any possible letter of the alphabet.

11 For an overview of the relationship between Morley and Shakespeare, see Long 1950.

12 Nick Bottom was probably played by William Kempe, a skilful dancer (as seen in Part I). This increases the likelihood that the Bergomask, with its high jumps, was meant for him.

13 According to Jackson and Neil, Bianca's Brawl is "possibly the name of an actual dance, or perhaps simply coined by Guerrino as an occasion to bawdy mockery of Bianca" (1986, 252).

14 As stated in Part I, Henrietta Maria was the youngest daughter of Henry IV of France and Maria de' Medici.

15 Issues surrounding this tragedy's authorship and date of composition are complex. It was probably written between 1612 and 1624, performed at the Globe in 1633 and at Court in 1637. It was recorded in the Stationer's Register in 1639 and published in a quarto version the same year. Fletcher, Chapman, Daborne, Field, Jonson, and Massinger have been suggested as possible authors/revisers by the critics. For detailed information, see Taylor and Jowett 1993, 260–271.

16 *The Spanish Gypsy* is also the title of one of the Country Dances listed in Playford's *The English Dancing Master*.

17 For the complex issues surrounding the authorship of this tragicomedy, see Taylor and Lavagnino 2007, 1723–27; Guardamagna 2018, 103–106.

18 Sinkapace is also the speaking name given to the dancing master in Thomas Middleton's *More Dissemblers Besides Women*.

19 Given the temper of the dancing master, one may assume that Frisk is a speaking name, since the lemma means "Skip or leap playfully; frolic" (*OED*, v. 2) or "A playful skip or leap" (*OED*, n. 2).

20 The reference to a Country Dance in *Misogonus* is considered to be the first in the entire corpus of early modern plays and in English in general, as attested in the *OED*.

21 Frankly, the consideration of Italy during the early modern period in England is much more complex and multifaceted. It is true that, on the one hand, Italy was the birthplace of any sin or vice, since Rome was where the Catholic Pope lived. Nonetheless, on the other hand, Italy was also considered the source of inspiration of the English Petrarchan poets, Neoplatonic philosophers, and peerless Renaissance artists. The literature on such a controversial topic abounds (see, i.a., Einstein 1902; Hale 1954; Fox 2011).

22 Nevertheless, contemporary race studies tend to correct such anti-presentist positions as Spiller's in favour of a general tendency to retrofit our understanding of racial differences and racism to the early modern period. See, i.a., Chapman 2017 for further details.

23 Anne Daye, director of Education and Research for The Historical Dance Society, prefers to use the adjective 'vernacular' (2017, online).

24 For the adherence of *A Woman Killed with Kindness* to the genre of the domestic tragedy, see Hopkins 1994.
25 The frequency of 'dance' on EEBO has been explored using the BYU corpora, created by Professor Mark Davies, available at https://www.english-corpora.org/eebo/.
26 Raw or absolute frequency is not statistically significant. In fact, relative frequency enables one to compare different texts according to the occurrences of some lexemes, words, parts of speech, or multiwords. It is obtained by dividing the raw frequency by the number of tokens a text has and multiplying the result by a common base (e.g., 10,000 or 1,000,000). Relative frequency is used to compare corpora or subcorpora of different sizes.
27 Note that Aldous Huxley used Marlowe's phrase 'antic hay' as title for his 1923 novel.
28 This performance of the Hay by devils may also recall, I would argue, the devils' dance in Marlowe's *Dr. Faustus* (1588), 2.1.
29 Many of the lemmas introduced in this part of the book occur in the multiword 'to dance + article + (premodifier) + name of the dance'. This happens, I would argue, for at least two reasons: firstly, some dance names of Italian/French origin were still not lexicalised in the sixteenth and seventeenth centuries. Secondly, in the case of dances whose signifier is polysemous, the above-mentioned multiword is needed to fix the meaning of 'dance' within a sentence.
30 In addition, when robbed by Andrew, Slipper promises the Antics he will find them again and "teach them to caper in a halter", thus probably implying that jumps were part of the choreography of the Hornpipe they have just performed with him.
31 This is not the only occurrence of the collocation 'lusty Hornpipe', this noun phrase also appearing in Heywood and Brome's *The Late Lancashire Witches* (1634), 3.3
32 In the play, Caponi is actually one of the three servants of Carolo Charomonte, tutor of Cosimo's nephew Giovanni de' Medici and father to Lidia.
33 For a thorough analysis of the occurrences of dances in Brome's canon as a metaphor for sexual innuendo and intercourse, see Paravano 2018, 27–28.
34 For a thorough understanding of the performance of the Jig after early modern plays, see Clegg 2019.
35 In this instance, the #Lancsbox KWIC function has been invaluable because I was struck by the fact that sometimes JIG collocated with such words as 'after', 'end', or 'done', thus arousing my curiosity. That is how I obtained the results discussed above.
36 Rhoads states that the phrase "stately measures" might have been taken from Dekker's 1608 tract *The Belman of London* (1936, 195).
37 For detailed explorations of the relationship between Neoplatonism and Catholicism in the Caroline era, see, i.a., Veevers 1989 and Tosi 2011.
38 The scene under scrutiny here has been repeatedly attributed to Middleton's revision of Shakespeare's original tragedy in 1616. See Inga-Stina Ewbank's introduction to this play in Taylor and Lavagnino's 2007 edition of Middleton's collected works (1165–1166) and issues related to Middleton's "new canon" in Guardamagna 2018 (233–237).

References

Anglo, S. (2007) The barriers: From combat to dance (almost). *Dance Research: The Journal of the Society for Dance Research* 25(2), 91–106.

Baskervill, C.R. (1929) *The Elizabethan Jig*. Chicago, IL: University of Chicago Press.

Blamires, A. (2018) The dating and attribution of Alphonsus, *Emperor of Germany. Early Theatre: A Journal Associated with the Records of Early English Drama* 21(2), 111–119.

Brewer, E.C. (1894) *Dictionary of Phrase and Fable*. Cambridge: Cambridge University Press.

Butler, C. (ed.) (2007) Jasper Fisher's Fuimus Troe. [Online]. Available at https://extra.shu.ac.uk/emls/iemls/renplays/fuimustroes.htm. [Accessed on 17 September 2020].

Chapman, M. (2017) *Anti-Black Racism in Early Modern Drama: The Other "Other"*. London/New York: Routledge.

Ciambella, F. (2017a) Dalla parola al *dumb dance show*: *Twelfth Night* del Synetic Theater. In: Tempera, M. and Elam, K. (eds.) *Twelfth Night. Dal testo alla scena*. Bologna: ELIM, 205–220.

Ciambella, F. (2017b) *"There was a star danced". Danza e rivoluzione copernicana in Shakespeare*. Roma: Carocci.

Ciambella, F. (2020) Danza, lingua e potere: (s)cortesia ne *La dodicesima notte* di Shakespeare. *Linguae &* 19(2), 13–33.

Clark, I. (1983) Character and cosmos in Marston's *Malcontent*. *Modern Language Studies* 13(2), 80–96.

Clegg, R. (2019) "When the play is done, you shall have a Jig or dance of all treads": Danced endings on Shakespeare's stage. In: Mcculloch, L. and Shaw, B. (eds.) *The Oxford Handbook of Shakespeare and Dance*. Oxford/New York: Oxford University Press, 83–106.

Craik, T.W. (ed.) (1995) *The Arden Shakespeare King Henry V*. London: Arden Shakespeare.

Crewe, J. (2009) Disorderly love: Sodomy revisited in Marlowe's *Edward II*. Criticism 51 (3), 385–399.

Cutting, J. (2005) *History and the Morris Dance: A Look at Morris Dancing from Its Earliest Days until 1850*. Alton: Dance Books.

Daye, A. (2017) A history of Country Dancing, with an emphasis on the steps. [Online]. Available at https://colinhume.com/dehistory.htm. [Accessed on 31 October 2020].

Dobson, M. and Wells, S. (eds.) (2001) *The Oxford Companion to Shakespeare*. Oxford: Oxford University Press.

do Desterro, I. (1998) Homophobia in Marlowe's *Edward II*. *Florianópolis* 34, 105–112.

Einstein, L. (1902) *The Italian Renaissance in England*. New York: Columbia University Press.

Elze, K. (ed.) (1867) *George Chapman's Tragedy of Alphonsus, Emperor of Germany*. Leipzig: F.A. Brockhaus.

Eubanks Winkler, A. (2020) *Music, Dance, and Drama in Early Modern English Schools*. Cambridge: Cambridge University Press.

Firth, J.R. (1957) *Papers in Linguistics 1934–1951*. Oxford: Oxford University Press.

Forrest, J. (1999) *The History of Morris Dancing: 1458–1750*. Toronto: University of Toronto Press.

Fox, M. (2011) The Italianate Englishman: The Italian influence in Elizabethan literature. *International Research and Review: Journal of Phi Beta Delta Honor Society of International Scholars* 1(1), 40–48.

Grove, G. (1900) *A Dictionary of Music and Musicians*. Oxford: Oxford University Press.

Guardamagna, D. (2018) *Thomas Middleton, drammaturgo giacomiano. Il canone ritrovato*. Roma: Carocci.

Hale, J. (1954) *England and the Italian Renaissance: The Growth of Interest in its History and Art*. London: Faber and Faber.

Hiscock, A. (2018) "Come, now a roundel and a fairy song": Shakespeare's *A Midsummer Night's Dream* and the early modern invitation to the dance. *Cahiers Élisabéthains: A Journal of English Renaissance Studies* 97(1), 39–68.

Holland, P. (ed.) (1998) *William Shakespeare's A Midsummer Night's Dream*. Oxford: Oxford University Press.

Hopkins, L. (1994) The false domesticity of *A Woman Killed with Kindness*. *Connotations* 4(1), 1–7.

Hopkins, L. (2016) *The Cultural Uses of the Caesars on the English Renaissance Stage*. London/New York: Routledge.

Hoskins, J. (2005) *The Dances of Shakespeare*. London/New York: Routledge.

Howard, S. (1993) Hands, feet, and bottoms: Decentering the cosmic dance in *A Midsummer Night's Dream*. *Shakespeare Quarterly* 44(3), 325–342.

Hutton, V. (1985) *A Midsummer Night's Dream*: Tragedy in comic disguise. *Studies in English Literature, 1500-1900* 25(2), 289–305.

Jackson, M.P. (2019) The date of *Alphonsus, Emperor of Germany*: The evidence of unique n-gram matches. *Notes and Queries* 66(4), 512–514.

Jackson, M.P. and Neil, M. (eds.) (1986) *The Selected Plays of John Marston*. Cambridge: Cambridge University Press.

Kelly, W.B. (1998) Mapping subjects in Marlowe's *Edward II*. *South Atlantic Review* 63(1), 1–19.

Lamb, M.E. (2013) Taken by the fairies: Fairy practices and the production of popular culture in *A Midsummer Night's Dream*. *Shakespeare Quarterly* 51(3), 277–312.

Long, J.H. (1950) Thomas Morley and Shakespeare. *Modern Language Notes* 65(1), 17–22.

McAlindon, T. (1973) *Shakespeare and Decorum*. London: Macmillan.

Mehl, D. (2011) *The Elizabethan Dumb Show: The History of a Dramatic Convention*. London/New York: Routledge.

Ostovich, H. (1994) "Teach you our princess English?": Equivocal translation of the French in *Henry V*. In: Trexlet, R. (ed.) *Gendering Rhetorics: Postures of Dominance and Submission in History*. Binghamton, NY: Medieval & Renaissance Texts & Studies, 147–162.

Paravano, C. (2018) *Performing Multilingualism on the Caroline Stage in the Plays of Richard Brome*. Newcastle upon Tyne: Cambridge Scholars Publishing.

Payne, I. (2003) *The Almain in Britain, c. 1549–c. 1675: A Dance Manual from Manuscript Sources*. Aldershot/Burlington, VT: Ashgate.

Pugliese, P.J. (2015) Parallels between fencing and dancing in late sixteenth century treatises. *Raymond J. Lord Collection of Historical Combat Treatises*. [Online]. Available at http://www.umass.edu/renaissance/lord/pdfs/Parallels.pdf. [Accessed on 24 September 2020].

Ravelhofer, B. (2006) *The Early Stuart Masque: Dance, Costume, and Music*. Oxford: Oxford University Press.

Rhoads, H.G. (ed.) (1936) *William Hawkins'* Apollo Shroving*: An Edition with Notes and an Introduction*. Philadelphia, PA: University of Pennsylvania Press.

Russel Brown, J. (1982) *Focus on Macbeth*. London: Routledge.

Sachs, C. (1937) *World History of the Dance*. New York: Norton.

Shaw, B. (2019) Shakespeare's dancing bodies: The case of Romeo. In: Mcculloch, L. and Shaw, B. (eds.) *The Oxford Handbook of Shakespeare and Dance*. Oxford/New York: Oxford University Press, 173–196.

Sorell, W. (1957) Shakespeare and the Dance. *Shakespeare Quarterly* 8(3): 367–384.

Spiller, E. (2011) *Reading and the History of Race in the Renaissance*. Cambridge: Cambridge University Press.

Taylor, G. and Jowett, J. (1993) *Shakespeare Reshaped, 1606–23*. Oxford: Oxford University Press.

Taylor, G. and Lavagnino, J. (eds.) (2007) *Thomas Middleton: The Collected Works*. Oxford: Oxford University Press.

Tosi, L. (2011) After Elizabeth: Representations of female rule in Massinger's tragicomedies. *Renaissance Studies* 25(3), 433–446.

Veevers, E. (1989) *Images of Love and Religion. Queen Henrietta and Court Entertainments*. Cambridge: Cambridge University Press.

Whitlock, K. (1999) John Playford's *The English Dancing Master* 1650/51 as cultural politics. *Folk Music Journal* 7(5), 548–578.

Wiggins, M. (2012) *British Drama 1533–1642: A Catalogue*. Vol. 8. Oxford: Oxford University Press.

Wiles, D. (2008) The carnivalesque in *A Midsummer Night's Dream*. In: Bloom, H. and Marson, J. (eds.) *A Midsummer Night's Dream*. New York: Bloom's Literary Criticism, 208–223.

Williams, G. (2001) Mediating between lexis and texts: collocational networks in specialised corpora. *ASp. La Revue de GERAS* 31–33, 63–76.

Winerock, E.F. (2005) Dance references in the records of early English drama: Alternative sources for non-courtly dancing, 1500–1650. *SDHS 2005 Proceedings: Twenty-eighth Annual Conference*. Evanston, IL: Northwestern University, 36–41.

Winerock, E.F. (2011) Staging dance in English Renaissance drama. *Dance Dramaturgy: Catalyst, Perspective, + Memory* (Proceedings of the thirty-fourth annual international conference of the Society of Dance History, York University and the University of Toronto, June 23–26, 2011). Toronto: Society of Dance History Scholars, 259–266.

Conclusion

The lexicographic analysis introduced in the first two parts of this book and carried out in the third offers a complex and multifaceted overview of the many connotations of dancing in early modern, pre-Playfordian English plays (1576–1651). Preliminary considerations about the practice of dance in sixteenth- and seventeenth-century England, on the one hand (Part I), and the tools offered by corpus linguistics, on the other (Part II), equally enabled efficient and rather rapid quantitative and qualitative analyses of the early modern English plays corpus created with the support of the VEP Early Modern Drama Collection and EEBO-TCP (666 texts in total, analysed using #Lancsbox in Part III).

First of all, the collocational patterning and lexicosemantic neighbourhood of the most important and widespread lexemes and lemmas concerning early modern English dances have been scrutinized via the various functions offered by the #Lancsbox software – i.e., GraphColl, KWIC, Welk, and Words. Next, the quantitative analysis sketched has been interpreted qualitatively in order to explore the many metaphoric, allegoric, and emblematic connotations attributed to dance which emerged from the results obtained by the corpus-informed collocational data extracted. In some cases, the outcomes found confirmed what has been stated by the most eminent critics about the various connotations of the different dances and choreographies within sixteenth- and seventeenth-century plays. In other cases, however, the data examined completed or partially contradicted the gender-oriented, social, ideological, and religious considerations outlined by early modern dance criticism, which in fact was the (hoped for) aim of this book. In other words, the corpus-based analysis presented in the previous chapters offered an ideal tool to enrich our understanding of early modern dancing through the exploration of recursive (or unique) collocational patternings of dance-related lexis.

Moreover, dance historians and other scholars researching sixteenth- and seventeenth-century dance in England have focused almost exclusively on Shakespeare's literary output (see, e.g., Sorell 1957; Brissenden 1981/2001; Hoskins 2005; McCulloch and Shaw 2019, as indicated in the references section of each part of this book). As Part III has demonstrated, the purpose of this book was to broaden this perspective by offering insights into the works of other Elizabethan, Jacobean, and Caroline playwrights who probably dealt with dancing more than Shakespeare. Undoubtedly, playwrights such as John Marston, Thomas Heywood, and James Shirley offer more detailed descriptions of how choreographies were actually performed during the sixteenth and seventeenth centuries in England – e.g., Guerrino's Brawl in Marston's *The Malcontent* or Monsieur Le Frisk's directions about posture and flow in Shirley's *The Ball.*

Gender- and sexuality-related issues, tensions between courtly and folk dances, and between Neoplatonic ideas and Puritan ethics clearly emerge from the analysis carried out. Such issues and tensions, as seen in Part III, seem to change slightly from one monarch to another – from the last Tudor queen to the first two Stuart kings – thus confirming a unity, a kind of *fil rouge* of themes and connotations attributable to the terpsichorean art.

In short, whether dancing in early modern England was deeply influenced by Italian and French Renaissance paradigms or struggled to find its own national dimension and dignity before Playford's *The English Dancing Master*, the connotations it acquired seem to insist on the same issues over the entire period examined here, regardless of which monarch was on the English throne.

Index

A
Abusch, D. 56
Acham, Roger 18
Acteon and Diana (Cox) 117
An Act for the Better Observation of the Lord's Day (1657) 21
The Alchemist (Jonson) 127
All's Well That Ends Well (Shakespeare) 88, 90
Alphonsus (Chapman and/or Peele) 129
The Anatomy of Abuses (Stubb(e)s) 30
Anglo, S. 80
The Annales of England (Stow) 17
ante litteram misogynism 30–31
anti-terpsichorean ideas 30–31
Antony and Cleopatra (Shakespeare) 109
Anzi, A. 40, 42
Apollo's Banquet (Playford) 112
Apollo Shroving (Hawkins) 117, 118, 126, 127
Apologie de la danse (de Lauze) 9, 10
An Apology for Poetry (Sidney) 27
Arbeau, Thoinot 2, 10, 16, 19, 84, 89
Arcades (Milton) 42
Aristotle 25
The Arraignment of Paris (Peele) 79, 108, 109
Ashton, Frederick 120
Asmole, Elias 13

B
Bach, J.S. 93
Baker, W.J. 31
Bakhitinian sense 43
Balanchine, George 84
The Ball (Shirley) 92, 93, 97, 99, 102, 138
Il ballarino (Fabritio Caroso) 2, 10, 11, 88
ballet's basic feet patterns 11
Bannerman, Henrietta 53
Barclay, Alexander 16
Barnaby Rich His Farewell to Military Profession (Rich) 16
Baskervill, C.R. 112
Beaumont, Francis 102, 121, 122
Beethoven, Ludwig van 51, 96
Bevington, D. 40
The Bloody Brothers or Rollo Duke of Normandy (Fletcher, *et al*) 89, 90
Blount, Thomas 62
Blow, John 20
Blurt, Master Constable; or the Spaniard's Night Walk (Dekker) 126, 127
body (or corporeal) turn 7
The Book Named the Governor (Elyot) 25, 29, 42
A Book of Notes and Common Places (Merbecke) 30
The Book of Sports (or *Declaration of Sports*) 21, 31
Brainard, Ingrid 11, 13
Braithwait(e), Richard 28
Brewer, E. Cobham 125
Brezina, Vaclav 67
The Brief Treatise, Concerning the

Use and Abuse of Dancing (Vermigli) 30
Brissenden, A. 1, 11, 15, 30, 31, 138
Britten, Benjamin 113
Brome, Richard 20, 97, 99, 100, 105, 108, 110, 112
Brown, M.L. 29
Brown, Russel 130
Buck (or Buc), Sir George 17
Buggins, Butler 14, 15
Bull, John 126
Bullokar, John 62
Byrd, William 124, 126

C
Caelum Britannicum (Carew) 20
Calendar of the Patent Rolls 15
Campaspe (Lyly) 128, 129
Campion, Thomas 102
Carew, Thomas 20
Caroline dances 1; reliable sources for study of 10
Caroline drama 122, 138
Caroline masques 20
Caroline period 14, 118
Castiglione, Baldassar 19, 24, 25, 27, 28, 82
Cavendish, Margaret 105, 112, 127
Cavendish, William 92, 104, 105, 114, 119, 120
Cawdrey, Robert 61, 62, 94
Caxton, William 2
Chapman, G. 80, 101, 129
Charles I, King 3, 8, 18, 20, 28, 38, 40, 41, 42
Charles II, King 2
Charnavel, Isabelle 56
Chloridia (Jonson) 102
Choreographics (Guest) 52
Chute, Marchette 15
Ciambella, Fabio 19, 29, 37, 53, 54, 118
The City Madam (Massinger) 122
The City Wit (Brome) 105
Civil War 2, 14, 15, 18, 31, 65, 96
Clark, I. 86
Cleland, James 27
Cockeram, Henry 62
cognitive semantics 5–56
cognitivism 55
Collier, John Payne 12
collocation in lexicography 128–129
The Comedy of Errors (Shakespeare) 15
communicative studies 77
Compan, Charles 59
Comus (Milton) 20
Concise Oxford Dictionary of Music 94, 96, 98, 99, 103, 117, 124, 126, 128
Conclusions upon Dances, both of This Age and the old (Lowin) 27
The Constant Maid (Shirley) 117
Coote, Edmund 62
Cornish, William 121
corpus-driven analysis 1, 49, 58–60
corpus linguistics 64–70
corruption from the promiscuous 30
Il Cortegiano (Castiglione) 19, 27, 82
Cotgrave, Randle 62
The Country Captain (W. Cavendish) 92
Couperin, François 88
The Court Beggar (Brome) 97, 99, 112
courtly dances 35–37; compared to folk dances 33–34; as courtly self-fashioning 35; human body as emblem of ideal Neoplatonic perfection in 35–36; mainly from France and Italy 36; manuscript sources 11–16; rigid and fixed schemes in 36; as worship of the devil 31
Covent Garden (Nabbes) 116
COVID-19 pandemic 65
Cowley, A. 94, 95, 96
Cox, Robert 117
Craik, T.W. 95
Crewe, J. 107
Cromwell, Oliver 2, 21, 28
Cropper, Dorothy N. 59
cultural studies 7
Cumber, John 91, 92
Cunningham, James P. 11
Cusano, Niccoló 24
The Cutter of Coleman Street (Cowley) 96
Cutting, J. 121

D
Dallington, Sir Robert 17
dance as courtship of women 25

dance as polysemic art 49
Dance Dictionary (Cropper) 59
dance ESP (English for Special/Specific Purposes) 58, 59, 60, 129–130
dance lexeme 101–102
Dance Nation Bureau (DNB), New York 52
Dance Notation (Guest) 52
dance-related dialogue 16
dance-related technical vocabulary 59–60
dances: Almain/Allemand(e) 10, 79–81; Argulius 14; Basse Danse 10, 16; The Beginning of the World 100; Bergomask 82–84; Black Almain 12; Branle/Brawl 9, 10, 84–87, 106; Canary 10, 88–91; Carol(e)s 129; Cinquepace/Sinkapace 91–93; Cinque Pas 16; continental 16; Coranto/Courante 9, 10, 93–96; Country Dance 36–37, 96–99, 102; Cross-dressed Dance 34; Cushion Dance 34, 99–100; Earl of Essex (or Earl of Essex Measure) 12; European country dances 97–98; female steps 9; Five Pas 14; French Branle 16; Furry Dance 33–34; Galliard 9, 10, 11, 16, 36, 38, 103–105, 112, 113, 125, 126; Gavotte 9; the Gigue 88; The Hay 100, 106–107, 125; the Hornpipe 16, 108–110; The Hunting of the Fox 100; the Jig/Gig/Gigue 16, 34, 88, 111–112; John Come Kiss Me Now 100; Les Bouffons 10; Lewd Dance 34; Long Dance 33; Madam Sosilia Almain (or Madam Cecilia Almain) 12; male steps 9; the Maypole Dance 35, 36, 109, 116–117, 129; Measure 117–119, 126; Moresca/Morisco 10, 119–120, 125; The Morris Dance 34–35, 36, 102, 119, 120–123, 129; Old Almain 12; Passamezzo 124–125, 126; Pavan 10, 36, 37, 118, 124, 125, 126–127; Quadran Pavan 12; Quadrille 96; Queen's Almain 12; Rogero 100; Rope Dance 34; Round (el)/Ring(let) 128–130; Saraband 36; Saxon Round 129–130; Scottish Jig 91, 92; The Shaking of the Sheets 100; Sir Roger de Coverley 96; Tinternell 12; Tom Tiler 100; Tourdion 10; Turkeylone 12; the Upspring 81; La Volta 10, 36, 39, 95, 113–115; Waltz 96; Welsh Morris Dance 123; Wooddicock 34
dance schools 15; royal monopoly on 15
The Dances of Shakespeare (Hoskin) 69, 70
dance writing 58–60
dancing as immoral 29–30
dancing bodies' bond 24–25
dancing relationship to politics 38–42
da Sermoneta, Marco Fabritio Caroso 2, 10, 11, 19, 88, 124
Davenant, William 41, 42
Davies, Sir John 26, 27, 29, 42
da Vinci, Leonardo 24
Davy, William (of Creed) 13
Day, John 86, 106, 107
Daye, Anne 11, 12
Dean-Smith, M. 21
de Chambonnières, Jacques Champion 88
Dekker, Thomas 83, 88, 89, 90, 103, 105, 110, 122, 126, 127
de Lauze, François 9, 10
Demmen, Jane 64
de Montagut, Barthélémy 9
Department of Dance, University of Iowa 54
Department of English, Opole University, Poland 58
Department of English, University of Toronto 62
Department of Linguistics, Harvard University 56
Desainliens, Claude 61
Desrat, Gustave 59
dialogism 53
Dialogue against Light, Lewd, and Lascivious Dancing (Fetherston) 30
A Dictionary French and English (Desainliens) 61
Dictionary of Phrase and Fable (Brewer) 125
A Dictionary of the English Language (Johnson) 2

Dictionnaire de la dance (Compan) 59
Dictionnaire de la danse (Desrat) 59
Digital Anthology of Early Modern English Drama (EMED) 65
Digital Humanities 3
Dispensatory (Renou) 62
Dizionario italiano De Mauro 125
Dobson, M. 125
do Desterro, I. 107
Dowland, John 126
Dudley, Robert 39
Duncan, Isadora 60

E
Early English Books Online (EEBO) repository 65, 66, 80, 82, 101, 137
Early Modern English Dictionaries Database (EMEDD) 62, 70
Early Modern English Lexicography (Schäfer) 61, 62
Early Modern English Plays (EMEP) corpus 67, 68, 69, 70, 80, 83, 85, 89, 98, 99, 100, 101, 102, 105, 106, 109, 110, 112, 114, 116, 118, 120, 125, 126, 127, 129, 137
early modern period: dance as emblem of harmony and order 41–42; dance discourse as controversial topic in England 23–24; definition 1–2, 64–65; Shakespeare's hegemonic position 64–65
Edward I, King 81
Edward II (Marlowe) 106, 107, 130
Elam, K. 61
Elizabethan dances 1; as form of political propaganda 41; reliable sources for study of 10
Elizabethan drama 122, 129, 138
Elizabethan period 14, 118; definition 2; French-originated dances in 17
Elizabethan Religious Settlement (1559) 28
Elizabeth I, Queen 3, 8, 15, 17, 18, 28, 33, 34, 36–37, 38, 80, 96, 103, 104, 113, 122, 125, 138
Elton, W.R. 15
Elyot, Sir Thomas 25, 26, 29, 42
Elze, K. 81
Encyclopædia Britannica 70, 116
Encyclopedia of Shakespeare's Language Project (ESLP) 64
English country dances 2, 19, 20; nationalisation of 60
The English Dancing Master (Playford) 2, 9, 19–21, 26, 60, 96, 98, 100, 138
English for Special/Specific Purposes (ESP) 58
The English Gentleman (Braithwait (e)) 28
Essay on Education (Locke) 19
Eubanks Winkler, A. 15
Every Man Out of His Humour (Jonson) 111, 112
The Example (Shirley) 86

F
Fabritio Caroso; *see* da Sermoneta, Marco Fabritio Caroso
*The Fair Maid of the West, Part 2 (*Heywood) 97, 98–99, 119, 120
Fauré, Gabriel 126
The Fawn (Marston) 114
Fenner, Dudley 31
Fetherston, Christopher 30
Ficino, Marsilio 24
La Fille mal gardée (ballet, Ashton) 120
The first part of the Elementarie (Mulcaster) 62
Firth, J.R. 128
Fisher, Jasper 117, 118
Fitzwilliam Virginal Book 124
Flecknoe, Richard 98
Fletcher, John 89, 90, 122
Florio, John 11, 61, 62
Florio's *Italian-English Dictionary* 11, 61
Folger Shakespeare Library 65, 66
folk-dance movement of 20th century 96
folk dances 33–35; circular choreographies in 129; compared to courtly 33–34; lack of rigid and fixed schemes in 36; roots of in British folklore 36
folk/popular balls 16–17
Ford, John 89, 90, 103, 105
Forrest 121

The Fortunate Isles and Their Union (Jonson) 106, 107
Foucauldian discourse 49
Freeman, Arthur 12
Freeman, Janet Ing 12
Fregoso, Ottaviano 25
Frye, S. 39
Fuimus Troes. The True Trojans (Fisher) 117, 118
Fulgens and Lucrece (Medwall) 16

G
Garfield, John 62
Garzoni, Tommaso 17
gender issues 7–8, 28–29, 95, 138
gender studies 7
Glapthorne, Henry 117, 119
Gloriana (opera, Britten) 113
Google Books 58
Goose, N. 17
Gosson, Stephen 30
Gough, John 115
Graham, Martha 60
grammar of dance 51
Grammatik der Tanzkunst (Zorn) (*Grammar of the Art of Dancing* trans. Sheafe) 51
The Great Duke of Florence (Massinger) 108, 109
Greene, Robert 108, 109, 111, 112
Grove, G. 127
Guardamagna, D. 40, 42
The Guardian (Cowley) 94, 95
Guest, Ann Hutchinson 52, 60
Gunter, Edward 12
Gunter, Eliner 12

H
Hammond, Sandra N. 59
Hanna, Judith L. 52
Hardy, Thomas 35
Hausted, Peter 94, 95
Hawkins, W. 117, 118, 126, 127
Heaney, M. 35
Henrietta Maria, Queen 20, 40, 41
Henry III, King 81
Henry IV, King 17, 41
Henry IV, Part 2 (Shakespeare) 120
Henry V (Shakespeare) 94, 95, 114
Henry VII, King 2
Henry VIII, King 16, 25, 36, 121
Herbert, Edward, Lord of Cherbury 18, 19
Heywood, Thomas 20, 97, 98, 99, 100, 114, 119, 120, 138
Hiscock, A. 129
History of the World (Raleigh) 42
Hoby, Thomas 82
Holbrook, P. 40
Holeman, Robert 14
Holland, Peter 83, 84
An Holy Dance 14
Homer 26
The Honest Whore, Part I (Dekker & Middleton) 83, 88, 89, 90
Hood Philips, O. 15
Hope, Jonathan 66
Hopkins, Lisa 100, 115
Hoskins, Jim 69, 70, 83, 86, 89, 90, 91, 106, 118, 138
Hotson, L. 41
Howard, Skiles 35, 36, 38, 83
human body in dance 7
Humour Out of Breath (Day) 86
Hutton, V. 83

I
Iamartino, G. 61
immoral dances 33–34
Inns of Court 2, 51, 81, 126; dance as confidence guarantor in 19; dance manuals from 11–16; dances at 12; early modern playwrights recorded at 15; John Playford at 20; staging of early modern plays at 15–16; teaching dance at 17
The Insatiate Countess 16
Institution of a Young Noble Man (Cleland) 27
International Writing Program, University of Iowa 54
Interregnum 2
The Introductory to Write and to Pronounce French (Barclay) 16
Italian dance manuals in 16th- and 17th-century Britain 17

J
Jack Drum's Entertainment (Marston) 16, 86, 112

Jackendoff, R. 56
Jacobean dances 1; reliable sources for study of 10
Jacobean drama 122, 127, 138
Jacobean period 14, 118; plays in 40–41
James I, King 3, 8, 9, 11, 17, 18, 31, 36, 38, 40, 41, 42, 125
James IV (Greene) 111, 112
James VI, King 36
Job (Williams) 126
Johnson, Samuel 2
Jones, Inigo 20, 40, 41
Jonson, Ben 20, 40, 42, 65, 100, 102, 106, 107, 111, 112, 127
Julius Caesar (Shakespeare) 34

K
Kaeppler, A. 7
Kelly, W.B. 107
Kempe, William 34, 35, 121, 122
Kemp's Nine Days Wonder. Performed in a Dance from London to Norwick (Kempe) 34, 121, 122
King Lear (Shakespeare) 54
The Knight of the Burning Pestle (Beaumont) 121, 122
Kraus, Lisa 55, 56

L
Laban, Rudolf 51, 52
Labanotation (Guest) 52, 60
Labanotation (or kinetography) 51–52
The Lady of Pleasure (Shirley) 86, 87, 94, 95
The Lady's Privilege (Glapthorne) 117, 119
Lamb, M.E. 84
Lancashire, A. 35
Lancashire, Ian 62
Lancaster University corpus toolbox (#Lancsbox) 67–69, 82, 122, 127, 129, 137
language and dance 49; biological foundation of 55; communication as basis for language for dance 53; dance lexis approach 49; language about dance approach 58–60; language for dance approach 49, 51–56
Language for Special/Specific Purposes (LSP) 58
Language of Dance Centre (LOD), London 52
Lavagnino, J. 90
The Law Tricks (Day) 106, 107
Lee, Sir Sidney 18
Lerdhal, F. 56
lexemes/lemmas (dance-related) 1, 4, 62, 129–130; Italian 11
lexical analysis 64; collocation 128–129
lexicography, early modern English 61–63
Lexicons of Early Modern English (LEME) project 62–63, 70
Il Libro del Cortegiano (Castiglione) 24–25
Limon, Jerzy 40
Lindahl, Greg 70
Lingua, or the Cobat of the Tongue and the Five Senses for Superiority (Tomkis) 113, 115, 125
linguistics 7, 55
Locke, John 19
Lodge, E. 38
Lord Hay's Masque (Campion) 102
Louange de la danse (de Montagut) 9, 10
Louis XIV, King 84
Love Labour's Lost (Shakespeare) 85
Lowes, Henry 20
Lowin, John 27
Lyly, John 128, 129

M
macaronic English 17
Macbeth (Shakespeare) 109
Macbeth (Shakespeare & Middleton) 128, 129–130
Macinnes, A.I. 36
MacLean, S. 39
The Malcontent (Marston) 16, 86, 138
Mammers, Katherine 9
*The Manner of Dancing of Basse Dances after the Use of France (*Barclay) 16
Marlowe, Christopher 106, 107, 130
The Marriage of Oceanus and Britannia (Flecknoe) 98

Marston, John 15, 65, 85, 112, 114, 138
Martin, Richard 26
Martyr, Peter 30
Masque of Blackness (Jonson) 40
The Masque of the Inner Temple (Beaumont) 102
masques 40, 65, 102; Caroline 40–41; Jacobean 40, 107
Massinger, Philip 85, 86, 93, 108, 109, 122
A Match at Midnight (Rowley) 121, 122–123
McAlindon, T. 83
McBride, K.B. 29
McCulloch, L. 1, 23, 29, 138
McEnery, Tony 67
McManus, C. 11
McMillan, S. 39
measure lemma 118–119, 128–129
Medwall, Henry 16
Mehl, D. 80
Mercurius Rusticus 41
Messalina (Nathaniel) 115
A Method for Travel (Darlington) 17
Microcosmus, a Moral Masque (Nabbes) 92, 93
Middleton, Thomas 83, 88, 89, 90, 112, 122, 124, 125, 128, 129
A Midsummer Night's Dream (Shakespeare) 37, 82, 129
Milton, John 20, 42
Misogunus (Rudd) 96, 98
Montini, D. 61
Montrose, L.A. 39
More Dissemblers Besides Women (Middleton) 124, 125, 129
Morini, M. 61
Morley, Thomas 82
The Morris Federation 35
The Morris Ring 35
Morton, Thomas 31
Mozart, Wolfgang Amadeus 51, 96
Much Ado About Nothing (Shakespeare) 91, 92, 112
Mulcaster, Richard 62
musica universalis (dance of celestial spheres) 25–26, 27, 28, 125

N
Nabbes, Thomas 92, 93, 116
Napoli, Donna Jo 55, 56
Nashe, Thomas 98
Nathaniel, Richard 115
nationalism 17, 36
Negri, Cesare 2, 10, 11, 19, 124
Neoplatonism 3, 8, 49, 138; adherence to 25–26; fulcrum of dance defence 24; *vs.* Puritanism 23–24
neuroscience 7
New Historicism 2
New Measures 12
Nichols, J. 37, 39
Nicol, E.J. 21
Nobiltà di dame (Caroso) 10
Northbrooke, John 29
Northward Ho (Dekker & Webster) 110
No Wit, No Help Like a Woman's (Middleton) 112
Nuove inventioni di balli (prev. *Le gratie d'amore*) (Negri) 2, 11

O
O'Connor, S. 35
Odyssey (Homer) 26
Old Measures 12, 13, 14, 15, 37, 51, 81, 126
Orchésographie (Arbeau) 2, 16, 30
L' Orchésographie (Thoinot) 84
Orchestra, or a Poem of Dancing (Davies) 26, 29, 42
Order for Suppressing Jigs at the End of Plays (Platter) 112
Osborne, Rowland 14
Osselton, N.E. 61
Ostovich, Helen 39, 95
Othello (Shakespeare) 120
Overthrow of Stage-Plays (Rainolds (or Reynolds)) 31
Oxford Companion to Shakespeare 70, 91
Oxford Dictionary of Dance 59, 71, 82, 84, 88, 106, 108, 111
Oxford Dictionary of English Etymology 94, 118, 119, 120, 127
Oxford Dictionary of Music 70, 79
Oxford English Dictionary (OED) 58, 59, 62, 85, 89, 101

The Oxford Handbook of Shakespeare and Dance (McCulloch & Shaw) 23

P
pagan rites 27–28
Pallavicino, Gaspare 25
Palsgrave, John 62
parametric typology of dance 55
Patel-Grosz, P. 55, 56
Pavane (Fauré) 126
Pavane pour une infant défunte (Ravel) 126
Payne, I. 81
Peasant dances 79
Peele, G. 79, 80, 108, 109, 129
Peggy Baker Dance Projects, Toronto, Canada 54
A Perfect Narrative of the Whole Proceedings of the High Court of Justice in the Trial of the King in Westminster Hall (Playford) 20
performance studies 7
Petrina, A. 2
Philips, Peter 124, 126
philosophy 7, 26; moral 25; Neoplatonic 23, 24, 49
La piazza universale di tutte le professioni del mondo (Garzoni) 17
Pico della Mirandola, Giovanni 24
The Picture (Massinger) 85, 86, 93
A Plain and Easy Introduction to Practical Music Set Down in Form of a Dialogue (Morley) 82
Plato 17, 24, 25, 26, 35
Platonism 25
Platter, Thomas 34, 112
Playford, John 2, 9, 19, 20, 26, 51, 60, 96, 98, 100, 112, 138
Podhajecka, Miroslawa 58
polysemic values (of dancing) 1
Prokoviev, Sergei 54
proto-Christian Platonic union of bodies 24
Pugliese, P.J. 80
Purcell, Henry 20
Puritanism 3, 8, 21, 49, 115, 138; closure of theatres in 1642 by 41; dance as reprehensible 24, 25; against dancing and sports on Sunday afternoons 31; majority of folk dances banned by 33–35; moral dictates of on folk dances 33; *vs.* Neoplatonism 23–24; as opposition to legitimacy of terpsichorean practice 28–29
Pythagoras 26

Q
The Queen and Concubine (Brome) 108, 110
Queen Anna's New World of Words (Florio) 61
Queen Elizabeth I Dancing with Robert Dudley, Earl of Leicester (painting) 38–39

R
Rainolds (or Reynolds), John 31
Raleigh, Sir Walter 42
Ramsey, J. 13
The Rape of Lucrece (Heywood) 114
Ravel, Maurice 126
Ravelhofer, B. 102
the Reformation 42
religious and moral querelle 8, 23–24
Renaissance period 2; ballets and mummeries 119; European dance masters during 10; first collection of choreographies 10; influence of continental dancing masters during 11; Italian dance manuals 51; play and dance after sermon in Europe 31
Renou, J. 62
Restoration 14, 21, 35
Rich, Barnaby 16
The Rival Friends (Hausted) 94, 95
The Roaring Girl (Dekker & Middleton) 122
Romeo and Juliet (Prokofiev ballet) 54
Romeo and Juliet (Shakespeare) 37
Rowley, William 89, 90, 122, 123
Rudd, Anthony 96, 98

S
Salmacida Spolia (Davenant) 41, 42
Sándorfia, M. 61
Sappho and Phao (Lyly) 128, 129

Savov, Smilen Antonov 53–54
Saward, John 30
Schäfer, Jürgen 61, 62
The Schoolmaster (Ascham) 18
The School of Abuse (Gosson) 30
Schrifttanz (journal) 51
The Scottish History of James IV (Peele) 108, 109
semantic of visual narrative 56
semiotics 53–54
Shakespeare, William: *All's Well That Ends Well* 88, 90; *Antony and Cleopatra* 1po; *The Comedy of Errors* 15; *Henry IV, Part 2* 120; *Henry V* 94, 95, 114; *Julius Ceasar* 34; *King Lear* 54; *Love Labour's Lost* 85; *Macbeth* 109; *Macbeth* (with Middleton) 128, 129–130; *A Midsummer Night's Dream* 37, 82, 129; *Much Ado About Nothing* 91, 92, 112; *Othello* 120; *Romeo and Juliet* 37; *The Taming of the Shrew* 83; *The Tempest* 40; *Troilus and Cressida* 15, 42, 114; *Twelfth Night* 15, 16, 103, 104, 124, 125; *The Two Noble Kinsmen* (with John Fletcher) 122; *The Winter's Tale* 37
Shakespeare and the Dance (Cunningham) 11
Shaw, B. 1, 23, 29, 80, 138
Sheafe, Alfonso J. 51
Shirley, James 41, 86, 92, 93, 94, 97, 99, 102, 104, 105, 114, 117, 119, 120, 138
A Short and Profitable Treatise of Lawful and Unlawful Recreations (Fenner) 31
Sidney, Philip 27
Siemens, Raymond 62
Sirach 30
Sir Giles Goosecap (Chapman) 101
Smith, B.R. 34
social class in dance 8
Sorell, Walter 53, 84, 107, 138
The Spanish Gipsy (Ford, Dekker, Middleton & Rowley) 89
Spenser, Edmund 42
Spiller, Elizabeth 99
Stanhope, Sir John 38
stereotyping 29
Stokes, James 11, 13
storytelling 53; intermingling with dance 54
Stow(e), John 13, 17
The Strange Discovery (Gough) 114, 115
Stravinsky, Igor 84
structuralist idea of dance narrative 53
Stuart period 20, 21, 34; criticisms of dancing in 41–42; distinction between elitist and popular dances in 33; relationship between dance and politics in 38, 40–42
Stubb(e)s, Philip 30
The Sun's Darling (Dekker & Ford) 103, 105
Sutton, Julia 33
Symposium (Plato) 24
Synetic Theatre, Arlington, Virginia 54

T
A Table Alphabeticall (Cawdrey) 61, 94
Talbot, Gilbert 38
The Taming of the Shrew (Shakespeare) 83
Tancred and Gismund (Wilmot) 79, 80–81
Taylor, G. 90
The Tempest (Shakespeare) 40
terpsichorean art of dance 7, 11; Country Dance as emblem of British patriotism 37; influence of Italian and French balls on English panorama 17; Neoplatonism *vs.* Puritanism in 23–24; Puritanism opposition to legitimacy of 28–29; Stuart period 20
terpsichorean knowledge 1, 2
Tess of the D'Urbervilles (Hardy) 35
textual linguistics 58–60
The Third Book of the Complete Country Dancing-Master (Walsh) 123
The Third University of England (Buck) 17
Thomas, Thomas 62
Thomas, William 62

Thorp, Jennifer 11, 12
Tillyard, E.M.W. 7, 27, 42
Tomkis, Thomas 113, 115, 126
tonal music 56
The Tragedy of Alphonsus, Emperor of Germany (Chapman and/or Peele) 80, 81
Traheron, Bartholomew 62
transmedialisation of fiction into choreography 54
Treatise Against Dancing, Dicing, Plays and Interludes, with Other Idle Pastimes (Northbrooke) 29
Treccani Italian Dictionary 125
The Triumph of Love and Beauty (Cornish) 121
The Triumph of Peace (Shirley) 41
Troilus and Cressida (Shakespeare) 15, 42, 114
Tsikurishvili, Irina 54
Tsikurishvili, Paata 54
Tuberbill, John 13
Tudor period 28, 34; relationship between dance and politics in 38–39
Turner, William 62
Twelfth Night (Shakespeare) 15, 16, 103, 104, 124, 125
The Two Merry Milkmaids, or the Best Words Wear the Garland (Cumber) 91, 92
The Two Noble Kinsmen (Fletcher & Shakespeare) 122

U
UNESCO International Dance Council 53
The Unfortunate Traveller (Nashe) 98
universal grammar of dance 56
University of Lancaster 64, 66

V
The Variety (W. Cavendish & Shirley) 103–104, 105, 114, 119, 120
Vermigli, Pietro Martire 30
Vigon, John 62
Villiers, George, 1st Duke of Buckingham 9, 42
Visualizing English Print (VEP) Early Modern Drama Collection 3, 66, 68, 82, 137
Vuillier, G. 34

W
Wagner, A. 18
Wagner, Richard 51
Walsh, John 123
A War of Words (Florio) 61
Webster, John 110
Wells, S. 125
Whitlock, K. 20, 21, 100
Wiesner, Susan L. 58
Wiles, D. 83
Williams, G. 128, 129
Williams, Vaughan 126
Willoughby, John 13, 15
Wilmot, Robert 79, 80
Wilson, David R. 11
Winerock, Emily F. 9, 28, 33, 34, 35, 42, 87, 100, 107
Winkler, Eubanks 102
The Winter's Tale (Shakespeare) 37
The Witch of Edmonton (Rowley) 122
The Wit's Cabal, Part 1 (M. Cavendish) 105, 112, 127
A Woman Killed with Kindness (Heywood) 99–100
Wright, Frederick A. 59

Y
You Move (Guest) 52
Yuzurihara, Akkio 60

Z
Zampolli, A. 62
Zorn, Friedrich A. 51

For Product Safety Concerns and Information please contact our EU
representative GPSR@taylorandfrancis.com
Taylor & Francis Verlag GmbH, Kaufingerstraße 24, 80331 München, Germany

www.ingramcontent.com/pod-product-compliance
Ingram Content Group UK Ltd.
Pitfield, Milton Keynes, MK11 3LW, UK
UKHW021056080625
459435UK00003B/22

* 9 7 8 0 3 6 7 5 4 0 4 7 0 *